移动互联网开发技术丛书

Spring Redis 实战开发

微课视频版

郭煦 编著

清華大學出版社
北京

内容简介

本书基础理论和工程项目相结合，循序渐进地介绍了应用 Spring 开发 Redis 应用程序的方法和技术。全书共 9 章，分别介绍 Redis 基础、Spring 基础、Spring Redis Template、Spring 操作 Redis 缓存、Redis 基础应用、响应式 Redis、Redis 集群和 Redis 仓库等知识及一个综合案例，书中的每个知识点都有相应的案例代码。

本书主要面向广大从事 Spring 开发、Java Web 开发、大数据开发的专业人员。

图书在版编目(CIP)数据

Spring Redis 实战开发：微课视频版/郭煦编著. —北京：清华大学出版社，2024.3(2024.7 重印)
(移动互联网开发技术丛书)
ISBN 978-7-302-65597-8

Ⅰ. ①S… Ⅱ. ①郭… Ⅲ. ①JAVA 语言－程序设计 Ⅳ. ①TP312.8

中国国家版本馆 CIP 数据核字(2024)第 045857 号

责任编辑：陈景辉 张爱华
封面设计：刘 键
责任校对：申晓焕
责任印制：杨 艳

出版发行：清华大学出版社
网　　址：https://www.tup.com.cn，https://www.wqxuetang.com
地　　址：北京清华大学学研大厦 A 座　**邮　　编**：100084
社 总 机：010-83470000　**邮　　购**：010-62786544
投稿与读者服务：010-62776969，c-service@tup.tsinghua.edu.cn
质量反馈：010-62772015，zhiliang@tup.tsinghua.edu.cn
课件下载：https://www.tup.com.cn，010-83470236
印 装 者：三河市人民印务有限公司
经　　销：全国新华书店
开　　本：185mm×260mm　**印　张**：16.5　**字　　数**：402 千字
版　　次：2024 年 4 月第 1 版　**印　　次**：2024 年 7 月第 2 次印刷
印　　数：1501～2500
定　　价：69.90 元

产品编号：099227-01

前言

FOREWORD

Redis作为基于键值对的NoSQL数据库，具有高性能、数据结构丰富、持久化、高可用性、分布式等特性，同时Redis本身非常稳定，已经得到业界的广泛认可和使用。掌握Redis已经逐步成为开发人员的必备技能之一。

作为一个轻量级框架，Spring可以有效组织项目中的中间件，为开发人员提供统一接口。Spring在当今软件开发领域以强大的优势占据着主流地位。

本书内容基于Redis 7.0.2和Spring 5编写，覆盖了应用Spring开发Redis应用程序的方方面面，从Redis常用命令、基本数据类型、RedisTemplate等基础知识到发布-订阅、Redis流、流水线、响应式Redis、Redis集群和Redis仓库等高级主题。为帮助读者学习和掌握相关的开发技术，书中提供了大量的案例。这些案例主要以电商(或社交)网络为应用背景。

本书主要内容

本书以案例为基础，非常适合具备一定Java基础的读者学习。通过学习本书，读者可以掌握利用Spring开发Redis应用程序的方法和技术。

全书共分为两大部分，共有9章。

第一部分Sping和Redis基础篇，包括第1～4章。第1章Redis基础，包括Redis简介、Redis特性、Redis安装和Redis常用命令。第2章Spring基础，包括Spring IoC及基础案例、配置元数据、Spring AOP和AspectJ AOP注解开发。第3章Spring Redis Template，包括Java Redis客户端、创建Redis连接、Spring操作Redis的5种基础数据类型及HyperLogLog、Redis位图和序列化及反序列化等。第4章Spring操作Redis缓存，包括JdbcTemplate、Spring整合Redis缓存、Redis缓存优缺点、缓存雪崩与缓存穿透。

第二部分Redis高级应用篇，包括第5～9章。第5章Redis基础应用，包括发布-订阅、Redis流、流水线、事务与Lua和Geo。第6章响应式Redis，包括Reactor简介、响应式Redis基础、使用ReactiveStringRedisTemplate、响应式发布-订阅。第7章Redis集群，包括主从复制集群、哨兵模式集群、Redis分片集群。第8章Redis仓库，包括入门程序、对象映射基础、对象-哈希映射、键空间、辅助索引、查询、生存时间、持久化和Redis数据仓库集群。第9章为Redis应用的综合案例。

在学习过程中难免遇到困难和不解，建议读者不要纠结于某个细节，可以先跳过问题往后学习。通常来讲，通过不断深入学习，前面不懂和疑惑的地方慢慢也就理解了。学习程序设计，一定要多动手实践，如果在实践过程中遇到问题，建议多思考，认真分析问题发生的原因，并在问题解决后总结经验。

本书特色

（1）内容全面，综合性强。本书涵盖了利用Spring开发Redis应用程序的全部核心知识点，同时涉及了软件工程领域的其他相关技术，如Java程序设计、软件测试、数据结构、函数式编程等。

（2）案例丰富，注重实战。本书涵盖了Redis日常应用的各方面，案例以电商或社交网络为背景，具有很强的实用性。

（3）简明易懂，代码详尽。本书以Redis 7.0.2为基础，对开发环境的搭建和代码的操作步骤都配备了详细的解释。

配套资源

为便于教与学，本书配有源代码、微课视频、教学课件、教学大纲、教案、软件安装包。

（1）获取源代码、软件安装包和全书网址方式：先扫描本书封底的文泉云盘防盗码，再扫描下方二维码，即可获取。

源代码

软件安装包

全书网址

（2）其他配套资源可以扫描本书封底的“书圈”二维码，关注后回复本书书号即可下载。

读者对象

本书主要面向广大从事Spring开发、Java Web开发、大数据开发的专业人员。

限于作者水平和时间，书中难免存在疏漏之处，欢迎广大读者批评指正。

作者

2024年1月

目录

CONTENTS

第 1 章

Redis基础

1.1 Redis 简介

随着互联网的不断发展,传统的关系数据库在处理超大规模纯动态网站时已无法满足海量数据存储的需要。而非关系数据库因具有较高的读写性能,能很好地解决大规模数据集合和多重数据种类带来的挑战及大数据应用难题。

NoSQL(Not Only SQL)泛指非关系数据库。NoSQL 没有关系数据库的关系特性,存储数据没有固定的模式,也不保证关系数据库的事务特性。得益于上述特点,NoSQL 数据库非常容易扩展,并具有很高的读写性能,即使在大数据量下,读写性能表现依然优秀。

目前,常见的 NoSQL 数据库产品有 Redis、MongoDB、Memcached、Cassandra 等。其中,Redis(Remote Dictionary Server,远程字典服务)是一个开源的、使用 ANSI C 编写的、基于内存的键值型 NoSQL 数据库。Redis 以单线程方式执行命令,所有命令串行执行,因此可以保证 Redis 命令执行的原子性。从 Redis 6.0 开始使用多线程处理网络请求,仍然用单线程执行命令。

Redis 支持多种语言的 API(Application Programming Interface,应用程序接口),包括 Java、Python、PHP、C 和 C++等。

作为 NoSQL 数据库,Redis 具有很高的读写性能,适用于高并发的应用场景和持久化存储。Redis 和另一个非常流行的 NoSQL 数据库 Memcached 类似,但 Redis 支持的数据类型更多,包括字符串(String)、哈希(Hash)、列表(List)、集合(Set)、有序集合(Sorted Set)、位图(Bitmap)、超级日志(HyperLogLog)、地理空间数据(Geospatial,Geo)和流(Stream)。Redis 支持对这些类型执行原子操作。例如,追加字符串,增加哈希中的值,将元素推送到列表,计算集合交集、并集和差集,或者获得排序集合中排名最高的成员,等等。

此外,为了确保内存数据的安全性,Redis 会定期将数据持久化到磁盘。为提高可用性,Redis 还支持集群模式。

由于 Redis 支持多种数据类型并且拥有优异的读写性能,它在实时数据存储、数据缓存、流和消息传递等方面得到了广泛的应用。从 Redis 官方统计来看,有很多企业都在使用 Redis,如 GitHub、Snapchat、Craigslist、Stack Overflow 等,国内企业如新浪微博、腾讯、百度、搜狐、优酷土豆、美团、小米、唯品会等。可以说,熟练使用 Redis 已经成为开发人员的必备技能。

1.2 Redis 特性

Redis 的特性可概括为以下几方面：

（1）支持多种数据类型。

（2）高性能读写。为了获取最佳性能，Redis 会将所有数据都存放在内存中，它的读写速度是非常快的。而且，Redis 还可以定期将内存的数据利用快照和日志的形式保存到硬盘上，这样在发生断电或机器故障时，不会导致数据丢失。Redis 执行数据持久化的方式有 RDB（Redis Database，Redis 数据库）和 AOF（Append Only File，仅追加文件）。RDB 持久化是以指定的时间间隔执行数据集的时间点快照。AOF 持久化是记录服务器接收到的每个写入操作，然后可以在服务器启动时再次回放这些操作，重建原始数据集。

Redis 之所以具有优异的读写性能，主要得益于以下几点：

① 纯内存操作。一般都是简单的读写操作，线程占用的时间很少，时间的开销主要集中在 I/O（Input/Output，输入输出）上，所以读写速度快。

② 采用单线程模型。这样可以保证每个操作的原子性，也减少了线程的上下文切换和竞争。

③ 使用 I/O 多路复用模型。将数据库的开、关、读和写都转换为事件；Redis 采用事件分离器，效率比较高。

④ 高效的数据结构。整个 Redis 就是一个全局哈希，其时间复杂度是 $O(1)$，而且 Redis 会执行再哈希操作以防止因哈希冲突导致链表过长。并且，为防止一次性重新映射时数据过大导致线程阻塞，Redis 采用了渐进式再哈希，巧妙地将一次性复制分摊到多次操作中，从而避免了阻塞，并加快了读写速度。Redis 对数据存储进行了优化，对数据进行压缩存储，还可以根据实际存储的数据类型选择不同的编码。

（3）可编程。Redis 提供了一个编程接口，允许用户在服务器上执行自定义脚本。在 Redis 7.0 及更高版本中，用户可以使用 Redis 函数来管理和运行脚本。在 Redis 6.2 及以下版本中，用户可以使用 Lua 脚本和 EVAL 命令对服务器进行编程。

（4）可扩展。Redis 提供了一个自定义扩展模块，支持使用 C、C++或 Rust 语言自定义 Redis 命令。除了支持功能扩展外，Redis 通过基于哈希的分片支持水平的集群扩展，在集群增长时可自动重新分区，可以扩展到数百万个节点。

（5）高可用。Redis 具有针对独立和群集部署的自动故障切换复制机制。在 Redis 复制机制的基础上（不包括 Redis 集群或 Redis 哨兵作为附加层提供的高可用功能），有一个易于配置和使用的主从复制机制。它允许副本 Redis 是主实例的精确副本。复制副本会在每次连接断开时自动重新连接到主服务器，并且无论主服务器发生什么情况，复制副本都会尝试成为它的精确副本。

1.3 Redis 安装

Redis 的版本号采用标准惯例：主版本号.副版本号.补丁级别。偶数的副版本号表示稳定版本，如 1.2、2.0、2.4 等；奇数的副版本号用来表示非稳定版本，如 Redis 2.9.x 是非稳定版本，而 Redis 3.0 是稳定版本。本书采用的 Redis 版本是 7.0.2。

1.3.1 在 Linux 上安装 Redis

Redis 能够兼容大部分的 POSIX（Portable Operating System Interface of UNIX，UNIX 可移植操作系统接口）系统，如 Linux、macOS X、OpenBSD 等。其中，比较典型的是 Linux 操作系统（如 CentOS、RedHat、Ubuntu、Debian 等）。在 Linux 系统上安装 Redis 通常有两种方式：第一种方式是通过各个操作系统的管理软件进行安装，如 Ubuntu/Debian 有 APT，Linux 的主要发行版都有 Snapcraft 工具；第二种方式是通过源代码执行安装，整个安装过程只需要 6 步，以安装 Redis 7.0.2 版本为例。

```
# 下载 Redis 指定版本的源代码到当前目录
$ wget https://download.redis.io/redis-7.0.2.tar.gz
# 解压缩 Redis 源代码压缩包
$ tar -xzvf redis-7.0.2.tar.gz
# 进入 Redis 目录
$ cd redis--7.0.2
# 编译(编译前确保操作系统已安装 GCC)
$ make
# 安装
$ make install
# 开启 Redis 服务器
$ redis-server
```

1.3.2 在 macOS 上安装 Redis

可借助 macOS 系统上的 Brew 工具安装 Redis。执行下述命令即可：

```
$ brew install redis
```

可利用下述命令开启 Redis 服务器：

```
$ redis-server
```

利用组合键 Ctrl+C 可终止 Redis 服务器运行。利用下述命令开启 Redis 客户端：

```
$ redis-cli
```

本书的全部代码已在 macOS Ventura 13.2、Redis 7.0.2 及 JDK 17.0.6 上调试通过。

1.3.3 在 Windows 上安装 Redis

Redis 官方不建议在 Windows 下使用 Redis，所以 Redis 官方网站没有 Windows 版本可以下载。读者可以下载由 Tomasz Poradowski 维护的开源 Windows 版 Redis 5.0（网址详见前言二维码）。

1.4 Redis 常用命令

本节介绍 Redis 的一些常用命令。学习它们可以为后续学习打下一个好的基础。Redis 的命令有上百个，如果纯靠背诵比较困难，但是如果理解了 Redis 的一些机制，会发现这些命令有很强的通用性。此外，Redis 的一些数据结构和命令必须在特定场景下使用，一且使用不当可能会对 Redis 本身或者应用程序造成致命伤害，因此，熟悉 Redis 的常用命令就显得至关重要。本节选取了 3.3～3.7 节中涉及的 Redis 用于操作 5 种基础数据类型的命令以及调试程序中常用的 Redis 命令进行介绍，要了解详细的 Redis 命令及其用法，可参考 Redis 官方网站。

1.4.1 键值相关命令

1. KEYS

命令格式：KEYS pattern

说明：KEYS * 匹配 Redis 中所有键(Key)；KEYS h?llo 匹配 hello、hallo 和 hxllo 等；KEYS h * llo 匹配 hllo 和 heeeeello 等；KEYS h[ae]llo 匹配 hello 和 hallo，但不匹配 hillo。返回值：符合给定模式的键列表。

例如，下面输入了 3 对字符串类型的键值对(注意，本书采用 redis >代表 Redis 客户端的命令行提示符，通常显示为 127.0.0.1:6379 >)：

```
redis > SET hello world
OK
redis > SET java jedis
OK
redis > SET python redis - py
OK
```

KEYS * 命令会将所有的键输出：

```
redis > KEYS *
1) "hello"
2) "java"
3) "python"
```

2. SCAN

命令格式：SCAN cursor [MATCH pattern] [COUNT count]

说明：SCAN 命令是一个基于游标的迭代器(Cursor Based Iterator)。SCAN 命令每次被调用之后，都会向用户返回一个新的游标，用户在下次迭代时需要使用这个新游标作为 SCAN 命令的游标参数，以此来延续之前的迭代过程。当 SCAN 命令的游标参数被设置为 0 时，服务器将开始一次新的迭代，而当服务器返回值为 0 的游标时，表示迭代已结束。

以下是一个 SCAN 的迭代过程示例：

```
redis > SCAN 0
1) "17"
2) 1) "key:12"
   2) "key:8"
   3) "key:4"
   4) "key:14"
   5) "key:16"
   6) "key:17"
   7) "key:15"
   8) "key:10"
   9) "key:3"
```

```
    10) "key:7"
    11) "key:1"
redis > SCAN 17
1) "0"
2) 1) "key:5"
   2) "key:18"
   3) "key:0"
   4) "key:2"
   5) "key:19"
   6) "key:13"
   7) "key:6"
   8) "key:9"
   9) "key:11"
```

在上面例子中，第一次迭代使用 0 作为游标，表示开始一次新的迭代。第二次迭代使用的是第一次迭代时返回的游标，即命令回复第一个元素的值：17。从上面的示例可以看到，SCAN 命令的回复是一个包含两个元素的数组，第一个数组元素是用于进行下一次迭代的新游标，而第二个数组元素则是一个数组，这个数组中包含了所有被迭代的元素。在第二次执行 SCAN 命令时，命令返回了游标 0，这表示迭代已经结束，整个数据集已经被完整遍历过了。以 0 作为游标开始一次新的迭代，一直调用 SCAN 命令，直到命令返回游标 0，这个过程称为一次完整遍历。

3. EXISTS

命令格式：EXISTS key

说明：检查给定键 key 是否存在。若键存在，则返回 1，否则返回 0。

例如：

```
redis > EXISTS java
(integer) 1
redis > EXISTS others
(integer) 0
```

4. DEL

命令格式：DEL key[key ...]

说明：删除给定的一个或多个键。不存在的键会被忽略。返回值：被删除键的数量。

例如：

```
redis > SET name huangz
OK
redis > DEL name
(integer) 1
```

5. TTL

命令格式：TTL key

说明：以秒为单位，返回给定键 key 的剩余生存时间(Time To Live，TTL)。返回值：当键不存在时，返回－2；当键存在但没有设置剩余生存时间时，返回－1；否则，以秒为单位，返回键的剩余生存时间。

6. EXPIRE

命令格式：EXPIRE key seconds

说明：为给定键 key 设置生存时间，当键过期时（生存时间为 0），它会被自动删除。返回值：当设置成功时返回 1。当键不存在或者不能为键设置生存时间时（如在低于 2.1.3 版本的 Redis 中更新键的生存时间），返回 0。

例如：

```
#key 存在,但没有设置剩余生存时间
redis> SET key value
OK
redis> TTL key
(integer) -1
#将 key 的生存时间设置为 30s
redis> EXPIRE key 30
(integer) 1
redis> TTL key
(integer) 28
```

7. MOVE

命令格式：MOVE key db

说明：将当前数据库的键 key 移动到给定的数据库 db 当中。如果当前数据库（源数据库）和给定数据库（目标数据库）有相同名字的键，或者键不存在于当前数据库，那么执行 MOVE 命令没有任何效果。返回值：若移动成功则返回 1，否则返回 0。

例如：

```
#key 存在于当前数据库
redis> SELECT 0              #Redis 默认使用数据库 0,为了清晰起见,这里再显式指定一次
OK
redis> SET song "secret base - Zone"
OK
redis> MOVE song 1           #将 song 移动到数据库 1
(integer) 1
redis> EXISTS song           #song 已经被移走
(integer) 0
redis> SELECT 1              #使用数据库 1
OK
redis:1> EXISTS song         #证实 song 被移到了数据库 1(注意,命令提示符变成了"redis:1",
                             #表明正在使用数据库 1)
(integer) 1
```

8. RANDOMKEY

命令格式：RANDOMKEY

说明：从当前数据库中随机返回（不删除）一个键。返回值：当数据库不为空时，返回一个键；当数据库为空时，返回 nil。

例如：

```
#数据库不为空
redis> MSET fruit "apple" drink "beer" food "cookies"   #设置多个键
OK
redis> RANDOMKEY
"fruit"
redis> RANDOMKEY
"food"
```

```
redis > KEYS *                # 查看数据库内所有键,证明 RANDOMKEY 并不删除键
1) "food"
2) "drink"
3) "fruit"
# 数据库为空
redis > FLUSHDB               # 删除当前数据库所有键
OK
redis > RANDOMKEY
(nil)
```

9. RENAME

命令格式：RENAME key newkey

说明：将键 key 改名为 newkey。当 key 和 newkey 相同，或者 key 不存在时，返回一个错误。当 newkey 已经存在时，RENAME 命令将覆盖旧值。返回值：改名成功时提示 OK，失败时候返回一个错误。

例如：

```
# key 存在且 newkey 不存在
redis > SET message "hello world"
OK
redis > RENAME message greeting
OK
redis > EXISTS message                    # message 不复存在
(integer) 0
redis > EXISTS greeting                   # greeting 取而代之
(integer) 1
# 当 key 不存在时,返回错误
redis > RENAME fake_key never_exists
(error) ERR no such key
# newkey 已存在时,RENAME 会覆盖旧 newkey
redis > SET pc "lenovo"
OK
redis > SET personal_computer "dell"
OK
redis > RENAME pc personal_computer
OK
redis > GET pc
(nil)
redis:1 > GET personal_computer          # 原来的值 dell 被覆盖了
"lenovo"
```

10. TYPE

命令格式：TYPE key

说明：返回键 key 所存储的值的类型。返回值：none（键不存在）、string（字符串）、list（列表）、set（集合）、zset（有序集合）、hash（哈希）。

例如：

```
# 字符串
redis > SET weather "sunny"
OK
redis > TYPE weather
string
```

```
#列表
redis > LPUSH book_list "programming in scala"
(integer) 1
redis > TYPE book_list
list
#集合
redis > SADD pat "dog"
(integer) 1
redis > TYPE pat
set
```

11. SORT

命令格式：SORT key[BY pattern][LIMIT offset count][GET pattern[GET pattern...]][ASC|DESC][ALPHA][STORE destination]

说明：返回或保存给定列表、集合、有序集合 key 中经过排序的元素。排序默认以数字作为对象，值可解释为双精度浮点数，然后进行比较。该命令最简单的用法是 SORT key 和 SORT key DESC：SORT key 返回键值从小到大排序的结果；SORT key DESC 返回键值从大到小排序的结果。因为 SORT 命令默认排序对象为数字，当需要对字符串进行排序时，需要显式地在 SORT 命令之后添加 ALPHA 修饰符，如：

```
#名称
redis > LPUSH name "rabbit "
(integer) 1
redis > LPUSH name "slight "
(integer) 2
redis > LPUSH name "investment"
(integer) 3
#按字符排序
redis > SORT name ALPHA
1) "investment"
2) "rabbit "
3) "slight"
```

排序之后返回元素的数量可以通过 LIMIT 修饰符进行限制，修饰符接受 offset 和 count 两个参数：offset 指定要跳过的元素数量；count 指定跳过 offset 个元素之后，要返回多少个对象，如：

```
#添加测试数据,列表值为 1～10
redis > RPUSH rank 1 3 5 7 9
(integer) 5
redis > RPUSH rank 2 4 6 8 10
(integer) 10
#返回列表中最小的 5 个值
redis > SORT rank LIMIT 0 5
1) "1"
2) "2"
3) "3"
4) "4"
5) "5"
```

可以使用外部键的数据作为权重，代替默认的直接对比键值的方式来进行排序。假设现在有用户数据如表 1-1 所示。

表 1-1　用户数据

uid	user_name_{uid}	user_level_{uid}
1	admin	9999
2	jack	10
3	peter	25
4	mary	70

将用户数据添加到 Redis 中，以 admin 为例：

```
#admin
redis > LPUSH uid 1
(integer) 1
redis > SET user_name_1 admin
OK
redis > SET user_level_1 9999
OK
```

默认情况下，SORT uid 直接按 uid 中的值排序，通过使用 BY 选项，可以让 uid 按其他键的元素来排序。例如，按照 user_level_{uid}的大小对 uid 排序：

```
redis > SORT uid BY user_level_ *
1) "2"        #jack , level = 10
2) "3"        #peter, level = 25
3) "4"        #mary, level = 70
4) "1"        #admin, level = 9999
```

其中，user_level_ * 是一个占位符，它先取出 uid 中的值，然后使用 user_level_1～user_level_4 的值作为排序 uid 的权值。

此外，使用 GET 选项，可以根据排序的结果取出相应的键值。如：

```
redis > SORT uid GET user_name_ *
1) "admin"
2) "jack"
3) "peter"
4) "mary"
```

还可以将该命令的 BY 选项和 GET 选项组合使用，读者可自行练习。

1.4.2　服务器相关命令

1. PING

命令格式：PING

说明：使用客户端向 Redis 服务器发送一个消息 PING，如果服务器运作正常，就会返回一个消息 PONG。该命令通常用于测试与服务器的连接是否仍然生效，或者用于测量延迟值。返回值：如果连接正常就返回一个 PONG，否则返回一个连接错误。

例如：

```
#客户端和服务器连接正常
redis > PING
PONG
#客户端和服务器连接不正常(网络不正常或服务器未能正常运行)
redis > PING
```

```
Could not connect to Redis at 127.0.0.1:6379: Connection refused
```

2. QUIT

命令格式：QUIT

说明：请求服务器关闭与当前客户端的连接。一旦所有等待中的回复(如果有的话)顺利写入客户端,连接就会被关闭。返回值：总是返回 OK(但是不会被打印显示,因为当时Redis-cli 已经退出)。

例如：

```
redis > QUIT
$
```

3. DBSIZE

命令格式：DBSIZE

说明：返回当前数据库中的键的数量。返回值：当前数据库的键的数量。

例如：

```
redis > DBSIZE
(integer) 5
redis > SET new_key "hello_moto"    # 增加一个键
OK
redis > DBSIZE
(integer) 6
```

4. FLUSHDB

命令格式：FLUSHDB

说明：清空当前数据库中的所有键。此命令从不失败。返回值：总是返回 OK。

例如：

```
redis > DBSIZE                       # 清空前数据库中的键的数量
(integer) 4
redis > FLUSHDB
OK
redis > DBSIZE                       # 清空后的键的数量
(integer) 0
```

5. FLUSHALL

命令格式：FLUSHALL

说明：清空整个 Redis 服务器的数据(删除所有数据库的所有键)。此命令从不失败。返回值：总是返回 OK。

例如：

```
redis > DBSIZE                       # 0 号数据库的键的数量
(integer) 9
redis > SELECT 1                     # 切换到 1 号数据库
OK
redis[1]> DBSIZE                     # 1 号数据库的键的数量
(integer) 6
redis[1]> flushall                   # 清空所有数据库的所有键
OK
redis[1]> DBSIZE                     # 不但 1 号数据库被清空了
```

```
(integer) 0
redis[1]> SELECT 0                    # 0号数据库(以及其他所有数据库)也一样被清空
OK
redis> DBSIZE
(integer) 0
```

1.4.3 字符串相关命令

1. SET

命令格式：SET key value[EX seconds][PX milliseconds][NX|XX]

说明：将字符串值 value 关联到键 key。如果键已经持有其他值，执行 SET 命令将覆盖其旧值，无视类型。对于某个原本带有生存时间的键来说，当 SET 命令在这个键上成功执行时，这个键原有的生存时间将被清除。关于可选参数，从 Redis 2.6.12 版本开始，SET 命令的行为可以通过一系列参数来修改：

EX seconds：设置键的过期时间，单位为秒。命令 SET key value EX seconds 的效果等同于命令 SETEX key second value。

PX milliseconds：设置键的过期时间，单位为毫秒。命令 SET key value PX milliseconds 的效果等同于命令 PSETEX key milliseconds value。

NX：只有在键不存在时，才对键进行设置操作。命令 SET key value NX 的效果等同于命令 SETNX key value。

XX：只有在键已经存在时，才对键进行设置操作。在 SET 命令设置操作成功完成时返回 OK。

例如：

```
#对不存在的键进行设置
redis> SET key "value"
OK
redis> GET key
"value"
#对已存在的键进行设置
redis> SET key "new-value"
OK
redis> GET key
"new-value"
#使用 EX 选项
redis> SET key-with-expire-time "hello" EX 10086
OK
redis> GET key-with-expire-time
"hello"
redis> TTL key-with-expire-time
(integer) 10069
#使用 PX 选项
redis> SET key-with-pexpire-time "moto" PX 123321
OK
redis> GET key-with-pexpire-time
"moto"
redis> PTTL key-with-pexpire-time
(integer) 111939
```

```
#使用 NX 选项
redis > SET not - exists - key "value" NX
OK                                          #键不存在,设置成功
redis > GET not - exists - key
"value"
redis > SET not - exists - key "new - value" NX
(nil)                                       #键已经存在,设置失败
redis > GET not - exists - key
"value"                                     #维持原值不变
#使用 XX 选项
redis > EXISTS exists - key
(integer) 0
redis > SET exists - key "value" XX
(nil)                                       #因为键不存在,设置失败
redis > SET exists - key "value"
OK                                          #先给键设置一个值
redis > SET exists - key "new - value" XX
OK                                          #设置新值成功
redis > GET exists - key
"new - value"
#NX 或 XX 可以和 EX 或者 PX 组合使用
redis > SET key - with - expire - and - NX "hello" EX 10086 NX
OK
redis > GET key - with - expire - and - NX
"hello"
redis > TTL key - with - expire - and - NX
(integer) 10063
redis > SET key - with - pexpire - and - XX "old value"
OK
redis > SET key - with - pexpire - and - XX "new value" PX 123321
OK
redis > GET key - with - pexpire - and - XX
"new value"
redis > PTTL key - with - pexpire - and - XX
(integer) 112999
#EX 和 PX 可以同时出现,但后面给出的选项会覆盖前面给出的选项
redis > SET key "value" EX 1000 PX 5000000
OK
redis > TTL key
(integer) 4993                              #这是 PX 参数设置的值
redis > SET another - key "value" PX 5000000 EX 1000
OK
redis > TTL another - key
(integer) 997                               #这是 EX 参数设置的值
```

2. GET

命令格式：GET key

说明：返回键 key 所关联的字符串值。假如键存储的值不是字符串类型,则返回一个错误,因为 GET 命令只能用于处理字符串值。返回值：若键不存在,则返回 nil；否则,返回键的值。如果键不是字符串类型,那么返回一个错误。

例如：

```
#对不存在的键或字符串类型键执行 GET 命令
```

```
redis> GET db
(nil)
redis> SET db redis
OK
redis> GET db
"redis"
```

3. GETSET

命令格式：GETSET key value

说明：将给定键 key 的值设为值 value，并返回键的旧值。当键存在但不是字符串类型时，返回一个错误。

例如：

```
redis> GETSET db mongodb                    #没有旧值，返回 nil
(nil)
redis> GET db
"mongodb"
redis> GETSET db redis                      #返回旧值 mongodb
"mongodb"
redis> GET db
"redis"
```

4. SETRANGE

命令格式：SETRANGE key offset value

说明：用参数 value 从偏移量 offset 开始覆盖给定键 key 所存储的字符串值。不存在的键当作空白字符串处理。SETRANGE 命令会确保字符串足够长以便将参数 value 设置在指定的偏移量上，如果键 key 原来存储的字符串长度比偏移量小(例如字符串只有 5 个字符长，但设置的 offset 是 10)，那么原字符串和偏移量之间的空白将用零字节("\x00")来填充。注意，能使用的最大偏移量是 $2^{29}-1$(536 870 911)，因为 Redis 字符串的大小被限制在 512MB 以内。如果需要使用更大的空间，则可以使用多个键。返回值：被 SETRANGE 修改之后字符串的长度。

例如：

```
#对非空字符串执行 SETRANGE 命令
redis> SET greeting "hello world"
OK
redis> SETRANGE greeting 6 "Redis"
(integer) 11
redis> GET greeting
"hello Redis"
#对空字符串/不存在的键执行 SETRANGE 命令
redis> EXISTS empty_string
(integer) 0
redis> SETRANGE empty_string 5 "Redis!"      #对不存在的键使用 SETRANGE
(integer) 11
redis> GET empty_string                      #空白处被"\x00"填充
"\x00\x00\x00\x00\x00Redis!"
```

5. GETRANGE

命令格式：GETRANGE key start end

说明：返回键 key 中字符串的子字符串，字符串的截取范围由 start 和 end 两个偏移量决定(包括 start 和 end 在内)。负数偏移量表示从字符串最后开始计数，-1 表示最后一个字符，-2 表示倒数第二个，以此类推。GETRANGE 命令通过保证子字符串的值域不超过实际字符串的值域来处理超出范围的值域请求。返回值：截取得出的子字符串。

例如：

```
redis > SET greeting "hello, my friend"
OK
redis > GETRANGE greeting 0 4                #返回索引 0~4 的字符,包括 4
"hello"
redis > GETRANGE greeting -1 -5              #不支持回绕操作
""
redis > GETRANGE greeting -3 -1              #负数索引
"end"
redis > GETRANGE greeting 0 -1               #从第一个到最后一个
"hello, my friend"
redis > GETRANGE greeting 0 1008611          #值域范围不超过实际字符串,超过部分自动被忽略
"hello, my friend"
```

6. MSET

命令格式：MSET key value[key value...]

说明：同时设置一个或多个键值对(key-value)。如果某个给定键已经存在，那么 MSET 命令会用新值覆盖旧值。如果需要在所有给定键都不存在的情况下执行设置操作，可考虑使用 MSETNX 命令。MSET 是一个原子性操作，所有给定键都会在同一时间内被设置。返回值：总是返回 OK(因为 MSET 不可能失败)。

例如：

```
redis > MSET date "2012.3.30" time "11:00 a.m." weather "sunny"
OK
redis > MGET date time weather
1) "2012.3.30"
2) "11:00 a.m."
3) "sunny"
```

7. MGET

命令格式：MGET key[key...]

说明：返回所有(一个或多个)给定键 key 的值。如果给定的键里面，有某个键不存在，那么返回 nil。因此，该命令永不失败。返回值：一个包含所有给定键的值的列表。

例如：

```
redis > SET redis redis.com
OK
redis > SET mongodb mongodb.org
OK
redis > MGET redis mongodb
1) "redis.com"
2) "mongodb.org"
redis > MGET redis mongodb mysql             #不存在的 mysql 返回 nil
1) "redis.com"
2) "mongodb.org"
```

```
3) (nil)
```

8. APPEND

命令格式：APPEND key value

说明：如果键 key 已经存在并且其值是一个字符串，APPEND 命令将值 value 追加到键原来的值的末尾。如果键不存在，APPEND 就简单地将给定的键设为值 value。返回值：追加值 value 之后键中字符串的长度。

例如：

```
#对不存在的键执行 APPEND
redis > EXISTS myphone                    #确保 myphone 不存在
(integer) 0
#对不存在的键进行 APPEND,等同于 SET myphone "nokia"
redis > APPEND myphone "nokia"
(integer) 5                               #字符长度
#对已存在的字符串进行 APPEND
redis > APPEND myphone " - 1110"          #长度从 5 个字符增加到 12 个字符
(integer) 12
redis > GET myphone
"nokia - 1110"
```

9. STRLEN

命令格式：STRLEN key

说明：返回键 key 所存储的字符串值的长度。当键存储的不是字符串值时，返回一个错误。

例如：

```
#获取字符串的长度
redis > SET mykey "Hello world"
OK
redis > STRLEN mykey
(integer) 11
#不存在的 key 长度为 0
redis > STRLEN nonexisting
(integer) 0
```

10. INCRBY/DECRBY

命令格式：INCRBY key delta/DECRBY key delta

说明：将键 key 所存储的值加上(或减少)增量 delta。如果键不存在，那么键的值会先被初始化为 0，再执行 INCRBY/DECRBY 命令。如果值包含错误的类型，或字符串类型的值不能表示为数字，那么返回一个错误。本操作的值限制在 64 位有符号数字表示的范围之内。

例如：

```
#对已存在的键进行 DECRBY
redis > SET count 100
OK
redis > DECRBY count 20
(integer) 80
#键不是数字值时
```

```
redis > SET book "long long ago..."
OK
redis > INCRBY book 200
(error) ERR value is not an integer or out of range
```

11. SETBIT

命令格式：SETBIT key offset value

说明：对键key所存储的字符串值，设置或清除指定偏移量offset上的位。位的设置或清除取决于value参数(1或0)。当键不存在时，自动生成一个新的字符串值。字符串会进行扩展以确保它可以将value保存在指定的偏移量上。当字符串值进行扩展时，空白位置以0填充。参数offset必须大于或等于0，小于2^{32}(位映射被限制在512MB之内)。返回值：指定偏移量原来存储的位。

12. GETBIT

命令格式：GETBIT key offset

说明：对键key所存储的字符串值，获取指定偏移量offset上的位。当偏移量比字符串值的长度大，或者键不存在时，返回0。返回值：字符串值指定偏移量上的位。

例如：

```
redis > set mykey 6
OK
redis > setbit mykey 7 1
(integer) 0
redis > getbit mykey 0
(integer) 0
redis > getbit mykey 6
(integer) 1
redis > get mykey
"7"
```

1.4.4 列表相关命令

1. LPUSH

命令格式：LPUSH key value[value...]

说明：将一个或多个值value插入列表key的表头。如果有多个值，那么各个值按从左到右的顺序依次插入表头。例如，对空列表mylist执行LPUSH mylist abc命令，列表的值将是cba，这等同于原子性地执行LPUSH mylist a、LPUSH mylist b和LPUSH mylist c三个命令。当key不存在时，会创建一个空列表并执行LPUSH命令。当key存在但不是列表类型时，返回一个错误。返回值：执行LPUSH命令后，列表的长度。

例如：

```
#加入单个元素
redis > LPUSH languages python
(integer) 1
#加入重复元素
redis > LPUSH languages python
(integer) 2
redis > LRANGE languages 0 -1                #列表允许重复元素
1) "python"
```

```
2) "python"
#加入多个元素
redis > LPUSH mylist a b c
(integer) 3
redis > LRANGE mylist 0 -1
1) "c"
2) "b"
3) "a"
```

2. RPUSH

命令格式：RPUSH key value[value...]

说明：将一个或多个值 value 插入列表 key 的表尾(最右边)。如果有多个值,那么各个值按从左到右的顺序依次插入表尾：例如对一个空列表 mylist 执行 RPUSH mylist a b c,得出的结果列表为 abc,等同于原子性地执行命令 RPUSH mylist a、RPUSH mylist b、RPUSH mylist c。当 key 不存在时,会创建一个空列表并执行 RPUSH 命令。当 key 存在但不是列表类型时,返回一个错误。返回值：执行 RPUSH 命令后,列表的长度。

例如：

```
#添加单个元素
redis > RPUSH languages c
(integer) 1
#添加多个元素
redis > RPUSH mylist a b c
(integer) 3
redis > LRANGE mylist 0 -1
1) "a"
2) "b"
3) "c"
```

3. LLEN

命令格式：LLEN key

说明：返回列表 key 的长度。如果 key 不存在,则返回 0。如果 key 不是列表类型,则返回一个错误。

例如：

```
#空列表
redis > LLEN job
(integer) 0
#非空列表
redis > LPUSH job "cook food"
(integer) 1
redis > LPUSH job "have lunch"
(integer) 2
redis > LLEN job
(integer) 2
```

4. LPOP

命令格式：LPOP key

说明：移除并返回列表 key 的头元素。当 key 不存在时,返回 nil。

例如：

```
redis > LLEN course
(integer) 0
redis > RPUSH course algorithm001
(integer) 1
redis > RPUSH course c++101
(integer) 2
redis > LPOP course                    #移除头元素
"algorithm001"
```

5. RPOP

命令格式：RPOP key

说明：移除并返回列表 key 的尾元素。当 key 不存在时，返回 nil。

例如：

```
redis > RPUSH mylist "one"
(integer) 1
redis > RPUSH mylist "two"
(integer) 2
redis > RPUSH mylist "three"
(integer) 3
redis > RPOP mylist                    #返回被弹出的元素
"three"
redis > LRANGE mylist 0 -1             #列表剩下的元素
1) "one"
2) "two"
```

6. LRANGE

命令格式：LRANGE key start stop

说明：返回列表 key 中指定区间[start，stop]内的元素。参数 start 和 stop 都以 0 表示列表的第一个元素，以 1 表示列表的第二个元素，以此类推。也可以用负数下标，以 −1 表示列表的最后一个元素，以 −2 表示列表的倒数第二个元素，以此类推。

例如：

```
redis > RPUSH fp-language lisp
(integer) 1
redis > LRANGE fp-language 0 0
1) "lisp"
redis > RPUSH fp-language scheme
(integer) 2
redis > LRANGE fp-language 0 1
1) "lisp"
2) "scheme"
```

7. LSET

命令格式：LSET key index value

说明：将列表 key 中下标为 index 的元素的值设置为 value，若操作成功则返回 OK。当参数 index 超出范围或对一个空列表(key 不存在)执行 LSET 命令时，返回一个错误。

例如：

```
#对空列表(key 不存在)执行 LSET 命令
redis > EXISTS list
```

```
(integer) 0
redis > LSET list 0 item
(error) ERR no such key
#对非空列表执行 LSET 命令
redis > LPUSH job "cook food"
(integer) 1
redis > LRANGE job 0 0
1) "cook food"
redis > LSET job 0 "play game"
OK
redis > LRANGE job 0 0
1)  "play game"
```

8. LTRIM

命令格式：LTRIM key start stop

说明：对一个列表进行修剪，让列表只保留指定区间[start，stop]内的元素，不在指定区间之内的元素都将被删除。参数 start 和 stop 都以 0 表示列表的第一个元素，以 1 表示列表的第二个元素，以此类推。也可以使用负数下标，以 −1 表示列表的最后一个元素，以 −2 表示列表的倒数第二个元素，以此类推。当 key 不是列表类型时，返回一个错误。

例如：

```
#情况 1: start 和 stop 都在列表的索引范围之内
redis > LRANGE alpha 0  - 1                    #alpha 是一个包含 5 个字符串的列表
1) "h"
2) "e"
3) "l"
4) "l"
5) "o"
redis > LTRIM alpha 1  - 1                     #删除 alpha 列表索引为 0 的元素
OK
redis > LRANGE alpha 0  - 1                    #"h"被删除了
1) "e"
2) "l"
3) "l"
4) "o"
#情况 2: stop 比列表的最大下标还要大
redis > LTRIM alpha 1 10086                    #保留 alpha 列表索引 1 至索引 10086 上的元素
OK
redis > LRANGE alpha 0  - 1                    #只有索引 0 上的元素"e"被删除了,其他元素还在
1) "l"
2) "l"
3) "o"
#情况 3: start 和 stop 都比列表的最大下标要大,并且 start < stop
redis > LTRIM alpha 10086 123321
OK
redis > LRANGE alpha 0  - 1                    #列表被清空
(empty list or set)
#情况 4: start 和 stop 都比列表的最大下标要大,并且 start > stop
redis > RPUSH new - alpha "h" "e" "l" "l" "o"  #重新建立一个新列表
(integer) 5
redis > LRANGE new - alpha 0  - 1
1) "h"
```

```
2) "e"
3) "l"
4) "l"
5) "o"
redis > LTRIM new - alpha 123321 10086          # 执行 LTRIM 命令
OK
redis > LRANGE new - alpha 0  - 1               # 同样被清空
(empty list or set)
```

9. LMOVE

命令格式：LMOVE source destination < LEFT | RIGHT >< LEFT | RIGHT >

说明：Redis 6.2 及以后版本支持此命令。该命令可执行下述原子性操作：返回并删除存储在源列表 source 的第一个或最后一个元素（头或尾，取决于 wherefrom 参数，取值为 LEFT 或 RIGHT），并将该元素推送到存储在目标列表 destination 中的第一个或最后一个位置（头或尾部，取决于 whereto 参数，取值为 LEFT 或 RIGHT）。

例如：

```
redis > RPUSH mylist "one" "two" "three"
(integer) 3
redis > LMOVE mylist myotherlist RIGHT LEFT
"three"
redis > LMOVE mylist myotherlist LEFT RIGHT
"one"
redis > LRANGE mylist 0  - 1
1) "two"
redis > LRANGE myotherlist 0  - 1
1) "three"
2) "one"
```

10. LINDEX

命令格式：LINDEX key index

说明：返回列表 key 中下标为 index 的元素。下标 index 以 0 表示列表的第一个元素，以 1 表示列表的第二个元素，以此类推。也可以使用负数下标，以－1 表示列表的最后一个元素，以－2 表示列表的倒数第二个元素，以此类推。如果 key 不是列表类型，则返回一个错误。

例如：

```
redis > LPUSH mylist "World" "Hello"
(integer) 2
redis > LINDEX mylist 0
"Hello"
redis > LINDEX mylist  - 1
"World"
```

11. LREM

命令格式：LREM key count value

说明：移除列表中与参数 value 相等的元素，数量为 count。count 的值可以是以下几种情况。

(1) count＞0：从表头开始向表尾搜索，移除与 value 相等的元素，数量为 count。

（2）count<0：从表尾开始向表头搜索，移除与 value 相等的元素，数量为 count 的绝对值。

（3）count=0：移除表中所有与 value 相等的值。

例如：

```
#先创建一个表 greet,内容排列如下
#morning hello morning hello morning
redis > LREM greet 2 morning                #移除从表头到表尾,最先发现的两个 morning
(integer) 2                                 #两个元素被移除
redis > LRANGE greet 0 2
1) "hello"
2) "hello"
3) "morning"
```

1.4.5　哈希相关命令

1. HSET

命令格式：HSET key field value

说明：将哈希 key 中的域 field 的值设为 value。如果 key 不存在，则创建一个新的哈希并执行 HSET 命令。如果域 field 已经存在于哈希中，则旧值将被覆盖。返回值：如果 field 是哈希中的一个新建域，并且值设置成功，则返回 1；如果哈希中域 field 已经存在且旧值已被新值覆盖，则返回 0。

例如：

```
redis > HSET website google "www.g.cn"          #设置一个新域
(integer) 1
redis > HSET website google www.google.com      #覆盖一个旧域
(integer) 0
```

2. HGET

命令格式：HGET key field

说明：返回哈希 key 中给定域 field 的值。当给定域不存在或是给定 key 不存在时，返回 nil。

例如：

```
#域存在
redis > HSET site redis redis.com
(integer) 1
redis > HGET site redis
"redis.com"
#域不存在
redis > HGET site mysql
(nil)
```

3. HMSET

命令格式：HMSET key field value[field value...]

说明：同时将多个域值对(field-value)设置到哈希 key 中。此命令会覆盖哈希中已存在的域。如果 key 不存在，则创建一个空哈希并执行 HMSET 命令。返回值：如果命令执行成功，则返回 OK；当 key 不是哈希类型时，返回一个错误。

例如：

```
redis> HMSET car byd Qin toyota HighLander
OK
redis> HGET car byd
"Qin"
redis> HGET car toyota
"HighLander"
```

4. HMGET

命令格式：HMGET key field[field...]

说明：返回哈希 key 中一个或多个给定域 field 的值。如果给定的域不存在于哈希中，那么返回一个 nil 值。因为不存在的 key 被当作一个空哈希来处理，所以对一个不存在的 key 执行 HMGET 命令将返回一个只带有 nil 值的表。返回值：一个包含多个给定域的关联值的表，值的排列顺序和给定域参数的请求顺序一样。

例如：

```
redis> HMSET pet dog "doudou" cat "nounou"     #一次设置多个域
OK
redis> HMGET pet dog cat fake_pet              #返回值的顺序和传入参数的顺序一样
1) "doudou"
2) "nounou"
3) (nil)                                       #不存在的域返回 nil 值
```

5. HGETALL

命令格式：HGETALL key

说明：返回哈希 key 中所有的域和值。在返回值里，紧跟每个域名之后是域的值，所以返回值的长度是哈希大小的两倍。

例如：

```
redis> HSET people jack "Jack Sparrow"
(integer) 1
redis> HSET people gump "Forrest Gump"
(integer) 1
redis> HGETALL people
1) "jack"                                      #域
2) "Jack Sparrow"                              #值
3) "gump"
4) "Forrest Gump"
```

6. HDEL

命令格式：HDEL key field[field...]

说明：删除哈希 key 中的一个或多个指定域 field，不存在的域将被忽略。返回值：被成功移除的域的数量，不包括被忽略的域。

例如：

```
#测试数据
redis> HGETALL abbr
1) "a"
2) "apple"
3) "b"
```

```
4) "banana"
5) "c"
6) "cat"
7) "d"
8) "dog"
#删除单个域
redis > HDEL abbr a
(integer) 1
#删除不存在的域
redis > HDEL abbr not - exists - field
(integer) 0
#删除多个域
redis > HDEL abbr b c
(integer) 2
redis > HGETALL abbr
1) "d"
2) "dog"
```

7. HLEN

命令格式：HLEN key

说明：返回哈希 key 中域的数量。当 key 不存在时，返回 0。

例如：

```
redis > HSET db redis redis.com
(integer) 1
redis > HSET db mysql mysql.com
(integer) 1
redis > HLEN db
(integer) 2
```

8. HEXISTS

命令格式：HEXISTS key field

说明：查看哈希 key 中，给定域 field 是否存在。返回值：如果哈希含有给定域，则返回 1；如果哈希不含有给定域，或 key 不存在，则返回 0。

例如：

```
redis > HEXISTS phone myphone
(integer) 0
redis > HSET phone myphone nokia - 1110
(integer) 1
redis > HEXISTS phone myphone
(integer) 1
```

9. HKEYS

命令格式：HKEYS key

说明：返回哈希 key 中的所有域。当 key 不存在时，返回一个空表。

例如：

```
#哈希非空
redis > HMSET book java Thinking in Java Python python programming
OK
redis > HKEYS book
```

```
1) "java"
2) "Python"
#空哈希或 key 不存在
redis > EXISTS fake_key
(integer) 0
redis > HKEYS fake_key
(empty list or set)
```

10. HVALS

命令格式：HVALS key

说明：返回哈希 key 中所有域的值。当 key 不存在时，返回一个空表。

例如：

```
#非空哈希
redis > HMSET book java Thinking in Java Python python programming
OK
redis > HVALS book
1) "Thinking in Java"
2) "python programming"
```

11. HSCAN

可参考 1.4.1 节 SCAN 命令。

12. HINCRBY

命令格式：HINCRBY key field increment

说明：为哈希 key 中的域 field 的值加上增量 increment。增量也可以为负数，相当于对给定域进行减法操作。如果 key 不存在，则创建一个新的哈希并执行 HINCRBY 命令。如果域 field 不存在，那么在执行命令前，域的值被初始化为 0。对一个存储字符串值的域 field 执行 HINCRBY 命令将报告错误。本操作的值被限制在 64 位有符号数字表示范围之内。

例如：

```
#increment 为正数
redis > HEXISTS counter page_view                #对空域进行设置
(integer) 0
redis > HINCRBY counter page_view 200
(integer) 200
redis > HGET counter page_view
"200"
#尝试对字符串值的域执行 HINCRBY 命令
redis > HSET myhash string hello,world           #设定一个字符串值
(integer) 1
redis > HINCRBY myhash string 1                  #命令执行失败，错误
(error) ERR hash value is not an integer
redis > HGET myhash string                       #原值不变
"hello,world"
```

1.4.6 集合相关命令

1. SADD

命令格式：SADD key member[member...]

说明：将一个或多个元素 member 加入集合 key 中，已经存在于集合的元素 member 将被忽略。假如 key 不存在，则创建一个只包含元素 member 作为成员的集合。当 key 不是集合类型时，返回一个错误。返回值：被添加到集合中的新元素的数量，不包括被忽略的元素。

例如：

```
#添加单个元素
redis > SADD bbs "discuz.net"
(integer) 1
#添加重复元素
redis > SADD bbs "discuz.net"
(integer) 0
#添加多个元素
redis > SADD bbs "tianya.cn" "groups.google.com"
(integer) 2
redis > SMEMBERS bbs
1) "discuz.net"
2) "groups.google.com"
3) "tianya.cn"
```

2. SMEMBERS

命令格式：SMEMBERS key

说明：返回集合 key 中的所有成员。不存在的 key 被视为空集合。

例如：

```
#key 不存在或集合为空
redis > EXISTS not_exists_key
(integer) 0
redis > SMEMBERS not_exists_key
(empty list or set)
#非空集合
redis > SADD language Ruby Python Clojure
(integer) 3
redis > SMEMBERS language
1) "Python"
2) "Ruby"
3) "Clojure"
```

3. SMOVE

命令格式：SMOVE source destination member

说明：将元素 member 从集合 source 移动到集合 destination。SMOVE 是原子性操作。如果集合 source 不存在或不包含指定的元素 member，则 SMOVE 命令不执行任何操作，仅返回 0；否则，元素 member 从集合 source 中被移除，并添加到集合 destination 中。当集合 destination 已经包含元素 member 时，SMOVE 命令只是简单地将集合 source 中的元素 member 删除。当 source 或 destination 不是集合类型时，返回一个错误。返回值：如果元素 member 被成功移除，则返回 1；如果元素 member 不是集合 source 的成员，并且没有对集合 destination 执行任何操作，则返回 0。

例如：

```
redis > SMEMBERS songs
```

```
1) "Billie Jean"
2) "Believe Me"
redis > SMEMBERS my_songs
(empty list or set)
redis > SMOVE songs my_songs "Believe Me"
(integer) 1
redis > SMEMBERS songs
1) "Billie Jean"
redis > SMEMBERS my_songs
1)  "Believe Me"
```

4. SREM

命令格式：SREM key member[member...]

说明：移除集合 key 中的一个或多个元素 member，不存在的元素 member 会被忽略。当 key 不是集合类型时，返回一个错误。返回值：被成功移除的元素的数量，不包括被忽略的元素。

例如：

```
#测试数据
redis > SMEMBERS languages
1) "c"
2) "lisp"
3) "python"
4) "ruby"
#移除单个元素
redis > SREM languages ruby
(integer) 1
#移除不存在元素
redis > SREM languages non - exists - language
(integer) 0
#移除多个元素
redis > SREM languages lisp python c
(integer) 3
redis > SMEMBERS languages
(empty list or set)
```

5. SPOP

命令格式：SPOP key

说明：随机移除并返回集合中的一个元素。如果只想随机获取一个元素，但不想该元素从集合中被移除，可以使用 SRANDMEMBER 命令。返回值：被随机移除的元素；当 key 不存在或 key 是空集时，返回 nil。

例如：

```
redis > SMEMBERS db
1) "MySQL"
2) "MongoDB"
3) "Redis"
redis > SPOP db
"Redis"
redis > SMEMBERS db
1) "MySQL"
2) "MongoDB"
```

6. SCARD

命令格式：SCARD key

说明：返回集合 key 的基数(集合中元素的数量)。当 key 不存在时，返回 0。

例如：

```
redis > SADD tool pc printer phone
(integer) 3
redis > SCARD tool                #非空集合
(integer) 3
redis > DEL tool
(integer) 1
redis > SCARD tool                #空集合
(integer) 0
```

7. SISMEMBER

命令格式：SISMEMBER key member

说明：判断元素 member 是否是集合 key 的成员。如果元素 member 是集合的成员，则返回 1。如果元素 member 不是集合的成员，或 key 不存在，则返回 0。

例如：

```
redis > SMEMBERS joe's_movies
1) "hi, lady"
2) "Fast Five"
3) "2012"
redis > SISMEMBER joe's_movies "bet man"
(integer) 0
redis > SISMEMBER joe's_movies "Fast Five"
(integer) 1
```

8. SSCAN

可参考 1.4.1 节 SCAN 命令。

9. SINTER

命令格式：SINTER key[key...]

说明：计算给定集合的交集。不存在的 key 被视为空集。返回值：交集成员的列表。

例如：

```
redis > SMEMBERS group_1
1) "LI LEI"
2) "TOM"
3) "JACK"
redis > SMEMBERS group_2
1) "HAN MEIMEI"
2) "JACK"
redis > SINTER group_1 group_2
1) "JACK"
```

10. SUNION

命令格式：SUNION key[key...]

说明：计算所有给定集合的并集。不存在的 key 被视为空集。返回值：并集成员的列表。

例如：

```
redis > SMEMBERS songs
1) "Billie Jean"
redis > SMEMBERS my_songs
1) "Believe Me"
redis > SUNION songs my_songs
1) "Billie Jean"
2) "Believe Me"
```

11. SDIFF

命令格式：SDIFF key[key...]

说明：计算所有给定集合之间的差集。不存在的 key 被视为空集。返回值：差集成员的列表。

例如：

```
redis > SMEMBERS peter's_movies
1) "bet man"
2) "start war"
3) "2012"
redis > SMEMBERS joe's_movies
1) "hi, lady"
2) "Fast Five"
3) "2012"
redis > SDIFF peter's_movies joe's_movies
1) "bet man"
2) "start war"
```

1.4.7 有序集合相关命令

1. ZADD

命令格式：ZADD key score member[[score member][score member]...]

说明：将一个或多个元素 member 及其分数 score 值加入有序集合 key 中。如果某个元素 member 已经是有序集合的成员，那么更新这个元素 member 的分数 score，并通过重新插入这个元素 member，来保证该元素 member 在正确的位置上。分数值可以是整数或双精度浮点数。如果 key 不存在，则创建一个空的有序集合并执行 ZADD 命令。当 key 存在但不是有序集合类型时，返回一个错误。返回值：被成功添加的新成员的数量，不包括那些被更新的、已经存在的成员。

例如：

```
#添加单个元素
redis > ZADD page_rank 10 google.com
(integer) 1
#添加多个元素
redis > ZADD page_rank 9 baidu.com 8 bing.com
(integer) 2
#添加已存在元素，且 score 值不变
redis > ZADD page_rank 10 google.com
(integer) 0
redis > ZRANGE page_rank 0 - 1 WITHSCORES              #没有改变
```

```
1) "bing.com"
2) "8"
3) "baidu.com"
4) "9"
5) "google.com"
6) "10"
#添加已存在元素,但是改变 score 值
redis > ZADD page_rank 6 bing.com
(integer) 0
redis > ZRANGE page_rank 0  -1 WITHSCORES              #bing.com 元素的 score 值被改变
1) "bing.com"
2) "6"
3) "baidu.com"
4) "9"
5) "google.com"
6) "10"
```

2. ZRANGE

命令格式：ZRANGE key start stop[WITHSCORES]

说明：返回有序集合 key 中，指定区间[start，end]内的成员。其中，成员的位置按 score 值升序排列。具有相同 score 值的成员按字典序排列。如果需要成员按 score 值降序排列，可使用 ZREVRANGE 命令。下标参数 start 和 stop 以 0 表示有序集合的第一个成员，以 1 表示有序集合的第二个成员，以此类推。也可以使用负数下标，以 −1 表示最后一个成员，以 −2 表示倒数第二个成员，以此类推。超出范围的下标并不会引起错误。例如，当 start 的值比有序集合的最大下标还要大或 start>stop 时，ZRANGE 命令只是返回一个空列表。另外，如果 stop 参数的值比有序集合的最大下标还要大，那么 Redis 将返回有序集合中的最后一个元素。可以使用 WITHSCORES 选项，让成员和它的 score 值一并返回，返回列表以 value1，score1，…，valueN，scoreN 的形式表示。

例如：

```
redis > ZRANGE salary 0  -1 WITHSCORES              #显示整个有序集合成员
1) "jack"
2) "3500"
3) "tom"
4) "5000"
5) "boss"
6) "10086"
redis > ZRANGE salary 1 2 WITHSCORES               #显示有序集合下标区间 1 至 2 的成员
1) "tom"
2) "5000"
3) "boss"
4) "10086"
redis > ZRANGE salary 0 200000 WITHSCORES          #测试 end 下标超出最大下标时的情况
1) "jack"
2) "3500"
3) "tom"
4) "5000"
5) "boss"
6) "10086"
#测试当给定区间不存在于有序集合时的情况
```

```
redis > ZRANGE salary 200000 3000000 WITHSCORES
(empty list or set)
```

3. ZREM

命令格式：ZREM key member[member...]

说明：移除有序集合 key 中的一个或多个成员，不存在的成员将被忽略。当 key 存在但不是有序集合类型时，返回一个错误。返回值：被成功移除的成员的数量，不包括被忽略的成员。

例如：

```
#测试数据
redis > ZRANGE page_rank 0  - 1 WITHSCORES
1) "bing.com"
2) "8"
3) "baidu.com"
4) "9"
5) "google.com"
6) "10"
#移除单个元素
redis > ZREM page_rank google.com
(integer) 1
#移除多个元素
redis > ZREM page_rank baidu.com bing.com
(integer) 2
redis > ZRANGE page_rank 0  - 1 WITHSCORES
(empty list or set)
#移除不存在元素
redis > ZREM page_rank non - exists - element
(integer) 0
```

4. ZCARD

命令格式：ZCARD key

说明：返回有序集合 key 的基数。当 key 不存在时，返回 0。

例如：

```
redis > ZADD salary 2000 tom                      #添加一个成员
(integer) 1
redis > ZCARD salary
(integer) 1
```

5. ZCOUNT

命令格式：ZCOUNT key min max

说明：返回有序集合 key 中分数值 score 在[min，max]区间内成员的数量。

例如：

```
redis > ZRANGE salary 0  - 1 WITHSCORES          #测试数据
1) "jack"
2) "2000"
3) "peter"
4) "3500"
5) "tom"
6) "5000"
```

```
redis > ZCOUNT salary 2000 5000                  # 计算薪水为 2000~5000 的人数
(integer) 3
redis > ZCOUNT salary 3000 5000                  # 计算薪水为 3000~5000 的人数
(integer) 2
```

6. ZSCORE

命令格式：ZSCORE key member

说明：以字符串形式返回有序集合 key 中元素 member 的 score 值。如果元素 member 不是有序集合 key 的成员或 key 不存在，则返回 nil。

例如：

```
redis > ZRANGE salary 0 - 1 WITHSCORES           # 测试数据
1) "tom"
2) "2000"
3) "peter"
4) "3500"
5) "jack"
6) "5000"
redis > ZSCORE salary peter                      # 注意返回值是字符串
"3500"
```

7. ZRANK

命令格式：ZRANK key member

说明：返回有序集合 key 中成员 member 的排名。其中，有序集合成员按分数值 score 升序排列。score 值最小的成员排名为 0。使用该命令可以获得成员按 score 值降序排列的排名。如果元素 member 不是有序集合 key 的成员，则返回 nil。

例如：

```
redis > ZRANGE salary 0 - 1 WITHSCORES           # 显示所有成员及其 score 值
1) "peter"
2) "3500"
3) "tom"
4) "4000"
5) "jack"
6) "5000"
redis > ZRANK salary tom                         # 显示 tom 的薪水排名，第二
(integer) 1
```

8. ZINCRBY

命令格式：ZINCRBY key increment member

说明：为有序集合 key 的元素 member 的分数值 score 加上增量 increment。增量 increment 可以为负值。当 key 不存在或元素 member 不是 key 的成员时，命令 ZINCRBY key increment member 等同于命令 ZADD key increment member。当 key 不是有序集合类型时，返回一个错误。分数值可以是整数或双精度浮点数。返回值：以字符串形式表示的元素 member 的新分数值。

例如：

```
redis > ZSCORE salary tom
"2000"
redis > ZINCRBY salary 2000 tom                  # tom 加薪了
"4000"
```

9. ZREVRANK

命令格式：ZREVRANK key member

说明：返回有序集合 key 中成员 member 的排名。其中，有序集合成员按分数值降序排列。分数值最大的成员排名为 0。如果 member 不是有序集合 key 的成员，则返回 nil。

例如：

```
redis 127.0.0.1:6379 > ZRANGE salary 0 - 1 WITHSCORES        # 测试数据
1) "jack"
2) "2000"
3) "peter"
4) "3500"
5) "tom"
6) "5000"
redis > ZREVRANK salary peter                                # peter 的工资排第二
(integer) 1
redis > ZREVRANK salary tom                                  # tom 的工资最高
(integer) 0
```

10. ZUNIONSTORE

命令格式：ZUNIONSTORE destination numkeys key[key...][WEIGHTS weight [weight...]][AGGREGATE SUM|MIN|MAX]

说明：计算给定的一个或多个有序集合的并集，其中参数 key 的数量必须以参数 numkeys 指定，并将该并集存储到集合 destination 中。默认情况下，结果集合中某个成员的分数值是所有给定集合下该成员分数值之和。使用 WEIGHTS 选项，用户可以为每个给定有序集合分别指定一个乘法因子，每个给定有序集合的成员的分数值在传递给聚合函数之前都要先乘以该有序集的乘法因子。如果没有指定 WEIGHTS 选项，则乘法因子取默认值 1。使用 AGGREGATE 选项，用户可以指定并集的结果集的聚合方式。默认使用的参数是 SUM，可以将所有给定集合中某个成员的分数值之和作为结果集中该成员的分数值；使用参数 MIN(MAX)，可以将所有给定集合中某个成员的最小(最大)分数值作为结果集中该成员的分数值。返回值：结果集 destination 的基数。

例如：

```
redis > ZRANGE programmer 0 - 1 WITHSCORES
1) "peter"
2) "2000"
3) "jack"
4) "3500"
5) "tom"
6) "5000"
redis > ZRANGE manager 0 - 1 WITHSCORES
1) "herry"
2) "2000"
3) "mary"
4) "3500"
5) "bob"
6) "4000"
 # 公司决定加薪，除了程序员
redis > ZUNIONSTORE salary 2 programmer manager WEIGHTS 1 3
(integer) 6
```

```
redis > ZRANGE salary 0 - 1 WITHSCORES
1) "peter"
2) "2000"
3) "jack"
4) "3500"
5) "tom"
6) "5000"
7) "herry"
8) "6000"
9) "mary"
10) "10500"
11) "bob"
12) "12000"
```

11. ZREMRANGEBYRANK

命令格式：ZREMRANGEBYRANK key start stop

说明：移除有序集合 key 中指定排名区间[start,stop]内的所有成员。区间分别以下标参数 start 和 stop 指出，以 0 表示有序集第一个成员，以 1 表示有序集第二个成员，以此类推。也可以使用负数下标，以−1 表示最后一个成员，以−2 表示倒数第二个成员，以此类推。返回值为被移除成员的数量。

例如：

```
redis > ZADD salary 2000 jack
(integer) 1
redis > ZADD salary 5000 tom
(integer) 1
redis > ZADD salary 3500 peter
(integer) 1
redis > ZREMRANGEBYRANK salary 0 1              # 移除下标 0 至 1 区间内的成员
(integer) 2
```

12. ZREVRANGEBYSCORE

命令格式：ZREVRANGEBYSCORE key max min [WITHSCORES][LIMIT offset count]

说明：返回有序集合 key 中分数值在[min,max]区间内的所有的成员。有序集成员按分数值降序排列。具有相同分数值的成员按字典序的逆序排列。

例如：

```
redis > ZADD salary 10086 jack
(integer) 1
redis > ZADD salary 5000 tom
(integer) 1
redis > ZADD salary 7500 peter
(integer) 1
redis > ZADD salary 3500 joe
(integer) 1
# 逆序排列薪水介于 10000 和 2000 之间的成员
redis > ZREVRANGEBYSCORE salary 10000 2000
1) "peter"
2) "tom"
3) "joe"
```

13. ZREVRANGEBYLEX

命令格式：ZREVRANGEBYLEX key max min[LIMIT offset count]

说明：当有序集合中的所有元素都以相同的分数插入时，为了强制进行反向字典排序，此命令将在键 key 处返回已排序集合的所有元素，其值在区间[min，max]内。从 Redis 6.2.0 版本开始，此命令被视为过时命令，可以用带有 REV 和 BYLEX 参数的 ZRANGE 命令替换它。

14. ZRANDMEMBER

命令格式：ZRANDMEMBER key[count[WITHSCORES]]

说明：当仅使用参数 key 时，从有序集合 key 中随机返回一个元素。如果提供的参数 count 为正，则返回一个含有不同元素的数组。数组的长度是 count 或排序集的基数，以较低者为准。如果参数 count 为负数，则允许命令多次返回同一元素。返回的元素个数是指定参数 count 的绝对值。可选的 WITHSCORES 参数，允许包含从有序集合中随机选择的元素的相应分数。Redis 6.2.0 及以后版本支持此命令。

例如：

```
redis> ZADD dadi 1 uno 2 due 3 tre 4 quattro 5 cinque 6 sei
(integer) 6
redis> ZRANDMEMBER dadi
"quattro"
redis> ZRANDMEMBER dadi -3 WITHSCORES
1) "cinque"
2) "5"
3) "sei"
4) "6"
5) "sei"
6) "6"
```

15. ZSCAN

可参考 1.4.1 节 SCAN 命令。

此外，还有很多 Redis 操作命令，篇幅所限，不再一一介绍。读者可查阅 Redis 官方网站了解相关命令的用法。

1.5 小结

Redis 是一款性能优秀的 NoSQL 数据库。Redis 将所有的数据都存放在内存中，因而具有良好的读写性能。而且，Redis 采用 C 语言编写，具有更高的执行效率。Redis 采用了单线程架构，从而避免了多线程可能产生的竞争问题，进一步提升了性能。从 Redis 6.0 开始，Redis 采用多线程处理网络请求，而对于读写命令仍然采用单线程。

鉴于其优秀的读写特性，Redis 得到了广泛的应用，如缓存、排行榜系统、社交网络、计数器（如视频网站的播放计数器、电商网站的浏览计数器等）、消息队列系统等。

学习 Redis，应从 Redis 提供的命令开始。由于 Redis 命令繁多，本章简要地将 Redis 的常用命令进行了归类，并着重介绍了后续章节中涉及的命令。

第 2 章

Spring基础

Spring 是一个轻量级的 Java 开发框架，目的是解决企业级应用开发的业务逻辑层和其他各层的耦合问题。Spring 设计良好，具有分层结构，克服了传统重量级框架臃肿、低效的劣势，极大降低了项目开发的复杂性。Spring 是一个开源框架，它集成了各种类型的工具。在整个框架中，各类型的功能组件被抽象成 Bean。通过核心的 Bean 工厂（Bean Factory）实现了组件（类）的生命周期管理。

作为一个轻量级开发框架，Spring 最核心的理念是控制反转（Inverse of Control，IoC）和面向切面编程（Aspect-Oriented Programming，AOP）。其中 IoC 是 Spring 的基础，支撑着 Spring 对 Bean 的管理功能；AOP 是 Spring 的重要特性，它通过预编译和运行时动态代理实现程序的功能。即在不修改源代码的情况下，为程序统一添加新的功能。本章针对 Spring 的两大特性 IoC 和 AOP 进行简单介绍。

2.1 Spring IoC

IoC 是 Spring 框架的核心，可以用来降低组件间的耦合度。依赖注入（Dependency Injection，DI）是 IoC 最常见的一种方式。对于 Spring 来说，控制反转和依赖注入只是从不同角度来描述同一个概念。下面通过日常生活中的例子来解释控制反转和依赖注入。

当人们需要某种东西时，第一反应是找东西。例如，需要面粉，在没有面粉厂时，需要人们自己将小麦磨成粉。如今，人们可以通过商家选择加工好的面粉。注意，这个时候人们没有自己动手磨制面粉，而是由面粉厂负责加工面粉，人们按照自己的需求购买。

上面举了一个非常简单的例子，但包含了控制反转和依赖注入的思想。即人们把磨制面粉的任务交给了厂家。就是说，当某个 Java 对象（调用者，例如面粉的消费者）需要调用另一个 Java 对象（被调用者，例如面粉）时，在传统的编程模式下，调用者会采用关键字 new 主动创建一个对象（消费者自己磨制面粉）。这种方式会增加调用者和被调用者之间的耦合度，不利于后期的升级与维护。

当 Spring 框架出现后，对象的实例不再由调用者创建（人们不再自己磨制面粉），而是由 Spring 容器负责创建（面粉厂生产面粉）。Spring 容器会负责控制对象之间的关系（将面粉送到消费者手中），而不需要由调用者负责控制。这样，控制权由调用者转换到 Spring 容器，这就是控制反转。

对于 Spring 容器而言，它负责将被调用对象赋值给调用者的属性，相当于为调用者注入它所依赖的对象，这就是Spring的依赖注入。在Spring中，依赖注入是指Spring容器在运行期间动态地将某依赖资源注入对象中。例如，将对象B注入（赋值）给对象A（的属性）。控制反转和依赖注入是从不同角度来描述同一件事情。控制反转是从应用程序的角度来描述，即应用程序将创建所需外部资源的权利交给了Spring；而依赖注入是从Spring容器的角度来描述，即Spring容器向应用程序注入其所需要的外部资源。

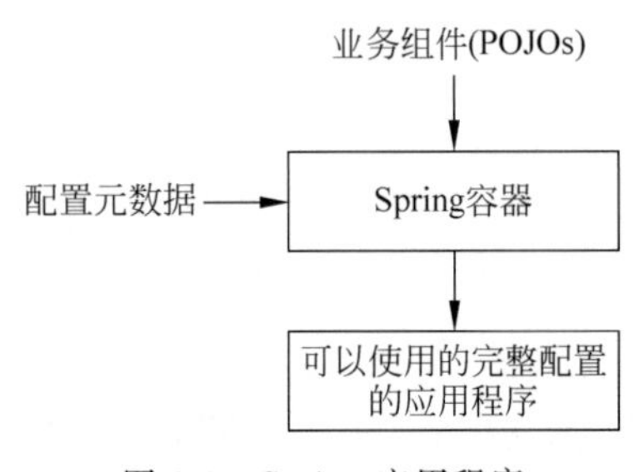

图 2-1　Spring 应用程序

综上所述，控制反转是一种通过声明（在Spring中可以是XML或注解）并借助第三方去产生或获取特定对象的方式。在Spring中实现控制反转的是Spring容器，其实现方法是依赖注入。为实现调用组件和被调用组件间的解耦，Spring将业务组件类与配置元数据相结合。这样，在创建和初始化Spring容器之后，就拥有了一个配置完整且可执行的应用程序，如图2-1所示。在Spring中，业务组件构成应用程序的主干，并由Spring容器管理，业务组件对象称为Bean。Bean是由Spring容器实例化、组装和管理的对象。Bean以及它们之间的依赖关系反映在Spring容器使用的配置元数据中。

2.2　配置元数据

Spring容器需要某种形式的配置元数据来实例化、配置和组装Bean。配置元数据可采用XML形式、Java注解形式和Java代码形式。Spring的配置元数据应至少包含一个Bean。本节只介绍基于Java注解形式的配置元数据。

使用注解形式的配置元数据一般采用@ComponentScan注解设置需要扫描的包，在指定的包中用注解标注Bean对应的类。其中，用于标注类的注解如表2-1所示。

表 2-1　Spring 常用配置元数据注解

注　解	说　明
@Component	用于描述Spring中的Bean，它是一个泛化概念，仅仅表示容器中的一个组件（Bean），并且可以作用在应用程序的任何层次，例如Service层、DAO层等。使用时只需将该注解标注在相应类上即可
@Repository	用于将数据访问（DAO）层的类标识为Spring中的Bean，其功能与@Component相同
@Service	用于将业务层（Service层）的类标识为Spring中的Bean，其功能与@Component相同
@Controller	用于将控制层（如SpringMVC的Controller）的类标识为Spring中的Bean，其功能与@Component相同
@Autowired	可以应用到Bean的属性变量、方法及构造方法，默认按照Bean的类型进行装配。默认情况下它要求依赖对象必须存在，如果允许null值，则可以设置它的required属性为false。如果想使用按照名称来装配，则可以结合@Qualifier注解一起使用
@Resource	作用与@Autowired注解相同，区别在于@Autowired默认按照Bean类型装配，而@Resource默认按照Bean的名称进行装配。@Resource中有两个重要属性：name和type。Spring将name属性解析为Bean的实例名称，type属性解析为Bean的实例类型。如果指定name属性，则按实例名称进行装配；如果指定type属性，则按Bean类型进行装配；如果都不指定，则先按Bean实例名称装配，如果不能匹配，则再按照Bean类型进行装配；如果都无法匹配，则抛出NoSuchBeanDefinitionException异常

续表

注　解	说　明
@Qualifier	与@Autowired 注解配合使用，会将默认的按 Bean 类型装配修改为按 Bean 的实例名称装配，Bean 的实例名称由@Qualifier 注解的参数指定
@Value	指定 Bean 实例的注入值
@Scope	指定 Bean 实例的作用域

提示：Spring 5 支持使用 JSR 250 中的@Resource 注解，即 javax. annotation. Resource。

利用注解方式指定配置元数据时，可在 com. example. spring. ioc 包中创建一个配置类 PersonConfig，代码如下：

```
1    @ComponentScan("com.example.spring.ioc.bean")
2    public class PersonConfig{
3    }
```

在 com. example. spring. ioc. bean 包及其子包中用@Component 指定 Bean 对应的类，如：

```
1    @Component
2    public class Chinese{
3      …  }
```

2.3 Spring IoC 基础案例

表 2-1 列举的注解是 Spring 常用的组件装配注解。需要注意的是，虽然@Controller、@Service 和@Repository 注解的功能与@Component 注解的功能相同，但为了使被标注的类本身的用途更加清晰，建议在开发中使用@Controller、@Service 和@Repository 注解分别标注控制器 Bean、业务逻辑 Bean 和数据访问 Bean。

下面，通过一个例子介绍 Spring IoC 的实现，步骤如下。

1. 创建项目并引入相关依赖

创建一个名为 myspring 的 Maven 项目，并引入以下依赖，其中 Spring 的版本为 5.3.18：

```
1          <dependency>
2              <groupId>junit</groupId>
3              <artifactId>junit</artifactId>
4              <version>4.12</version>
5              <scope>test</scope>
6          </dependency>
7          <!-- Spring 核心包 spring-beans -->
8          <dependency>
9              <groupId>org.springframework</groupId>
10             <artifactId>spring-beans</artifactId>
11             <version>${spring.version}</version>
12         </dependency>
13         <!-- Spring 核心包 spring-core -->
14         <dependency>
15             <groupId>org.springframework</groupId>
```

```
16                  <artifactId>spring-core</artifactId>
17                  <version>${spring.version}</version>
18              </dependency>
19              <!-- Spring 核心包 spring-context -->
20              <dependency>
21                  <groupId>org.springframework</groupId>
22                  <artifactId>spring-context</artifactId>
23                  <version>${spring.version}</version>
24              </dependency>
25              <!-- Spring 核心包 spring-expression -->
26              <dependency>
27                  <groupId>org.springframework</groupId>
28                  <artifactId>spring-expression</artifactId>
29                  <version>${spring.version}</version>
30              </dependency>
31              <dependency>
32                  <groupId>org.springframework</groupId>
33                  <artifactId>spring-test</artifactId>
34                  <version>${spring.version}</version>
35              </dependency>
36              <!--Spring 依赖包 commons-logging -->
37              <dependency>
38                  <groupId>commons-logging</groupId>
39                  <artifactId>commons-logging</artifactId>
40                  <version>1.2</version>
41              </dependency>
```

2. 编写DAO组件代码

在src/main/java目录下创建一个名为com.example.spring.ioc.dao的包，并在该包中创建DAO组件的接口和实现类，代码分别如文件2-1和文件2-2所示。

【文件2-1】 TeamDao.java

```
1   package com.example.spring.ioc.dao;
2   public interface TeamDao {
3       public void race();
4   }
```

【文件2-2】 TeamDaoImpl.java

```
1   package com.example.spring.ioc.dao;
2   import org.springframework.stereotype.Repository;
3
4   @Repository("teamDao")
5   public class TeamDaoImpl implements TeamDao {
6       public void race(){
7           System.out.println("dao: This is dao speaking.");
8       }
9   }
```

如文件2-2所示，DAO组件TeamDaoImpl由@Repository注解标注(第4行)。该注解的参数声明由Spring创建的Bean的名称为teamDao。

3. 编写 Service 组件代码

在 com. example. spring. ioc. service 包中创建 Service 组件的接口和对应的实现类，代码分别如文件 2-3 和文件 2-4 所示。

【文件 2-3】 TeamService. java

```
1    package com.example.spring.ioc.service;
2    public interface TeamService {
3        public void race();
4    }
```

【文件 2-4】 TeamServiceImpl. java

```
1    package com.example.spring.ioc.service;
2    //import 部分略
3    @Service("teamService")
4    public class TeamServiceImpl implements TeamService {
5
6        @Autowired
7        private TeamDao teamDao;
8
9        public void race(){
10           teamDao.race();
11           System.out.println("service: This is service speaking.");
12       }
13   }
```

如文件 2-4 所示，Service 组件 TeamServiceImpl 由@Service 注解标注(第 3 行)。该注解的参数声明由 Spring 创建的 Bean 的名称为 teamService。第 6 行使用@Autowired 注解实现 teamService 组件和 teamDao 组件的自动绑定。

4. 编写 Controller 组件代码

在 com. example. spring. ioc. controller 包中创建 Controller 组件，代码如文件 2-5 所示。

【文件 2-5】 TeamController. java

```
1    package com.example.spring.ioc.controller;
2    //import 部分略
3    @Controller("teamController")
4    public class TeamController {
5
6        @Autowired
7        private TeamService teamService;
8
9        public void race(){
10           teamService.race();
11           System.out.println("controller: That's all.");
12       }
13   }
```

如文件 2-5 所示，Controller 组件 TeamController 由@Controller 注解标注(第 3 行)。该注解的参数声明由 Spring 创建的 Bean 的名称为 teamController。第 6 行使用@Autowired 注解实现 teamController 组件和 teamService 组件的自动绑定。

5. 编写测试类

在 src/test/java 目录下,在 com.example.spring.test 包中创建测试类 TestSpringIoC,代码如文件 2-6 所示。

【文件 2-6】 TeamSpringIoC.java

```
1    package com.example.spring.test;
2    //import 部分略
3    @RunWith(SpringJUnit4ClassRunner.class)
4    @ContextConfiguration(classes = TeamConfig.class)
5    public class TestSpringIoC {
6
7        @Autowired
8        private TeamController tc;
9
10       @Test
11       public void testAssemblyByAnnotation(){
12           tc.race();
13       }
14   }
```

如文件 2-6 所示,第 3 行用 @RunWith 注解指定测试代码的运行器 SpringJUnit4ClassRunner,这个运行器可以在测试开始时自动创建 Spring 应用上下文,这样可以在测试类中加载 Spring 配置。第 4 行的@ContextConfiguration 注解是 Spring 整合 JUnit 4 测试时使用注解引入配置类,如加载配置类 TeamConfig。执行测试代码,可以在控制台得到三类组件完成组装调用的运行结果,如图 2-2 所示。

```
dao: This is dao speaking.
service: This is service speaking.
controller: That's all.
```

图 2-2 三类组件完成组装调用的运行结果

2.4 Spring AOP

2.4.1 AOP 概念

AOP 是 OOP(Object-Oriented Programming,面向对象编程)的延续,提供了与 OOP 不同的抽象软件结构的视角。要理解 AOP,首先要理解什么是横切关注点。在传统的业务处理代码中,通常有日志记录、性能统计、安全控制、事务处理等操作。虽然使用 OOP 可以通过封装或继承的方式实现代码的重用,但仍然会有同样的代码分散在各个方法中。这些散布于多处的功能被称为横切关注点(Cross-Cutting Concern)。例如,安全就是一个横切关注点,应用程序中的许多方法都会涉及安全规则。图 2-3 呈现了横切关注点的概念。这些横切关注点从概念上来说是与应用的业务逻辑相分离的,而 OOP 很难将横切关注点与业务逻辑分离,往往是直接嵌入应用的业务逻辑中。把这些横切关注点与业务逻辑分离正是 AOP 要解决的问题。AOP 采取横向抽取机制,将分散在各个方法中的重复代码提取出来,在程序编译或运行阶段将这些抽取出来的代码应用到需要执行的地方。这样做有两个好处:首先,每个关注点都集中于一个地方,而不是分散到多处代码中;其次,服务模块更简洁,因为它们只包含主要关注点(或核心功能)的代码,而次要关注点的代码被转移到切面(切面的概念见 2.4.2 节)中。这种横向抽取机制是 OOP 无法办到的,因为 OOP 实现的是父子关系的纵向重用,而 AOP 实现的是关注点的横向抽取。OOP 以类作为程序的基本单元,而 AOP 以切面作为程序的基本单元。

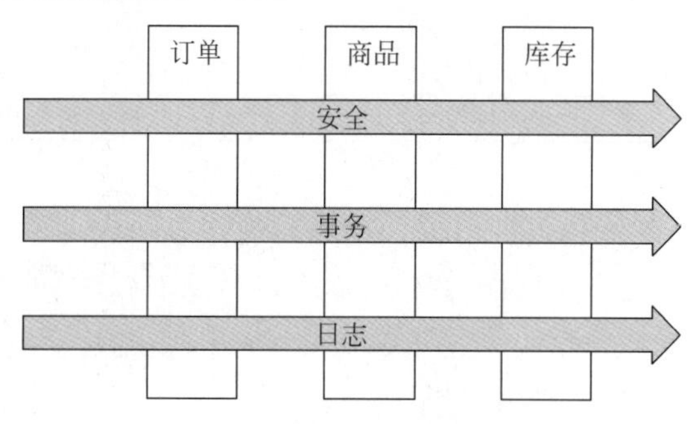

图 2-3　横切关注点的概念

2.4.2　AOP 术语

AOP 并不是一个新的概念。在 Java 中,早已出现了类似的机制。Java 平台的 EJB 规范、Servlet 规范和 Struts2 框架中的拦截器机制,与 AOP 的实现机制非常相似。AOP 是在这些概念基础上发展起来的。与大多数的技术一样,AOP 已经形成一套属于自己的概念和术语,图 2-4 简单解释了 AOP 中的一些术语。

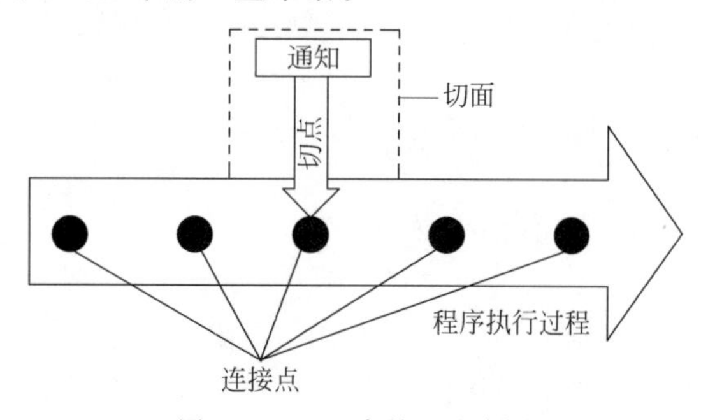

图 2-4　AOP 中的一些术语

1. 连接点

连接点(Joinpoint)是程序执行过程中的一个点,如方法调用或异常处理。在 Spring AOP 中,连接点总是表示方法的执行。切面代码可以借助这些点插入应用程序的正常流程中,并添加新的行为。

2. 切面

横切关注点(如事务管理、安全规则、日志记录等)可以被模块化为特殊的类,这些类被称为切面(Aspect)。

3. 通知

通知(Advice)是切面在连接点采取的操作,即通知定义了切面是什么以及何时使用。除了描述切面要完成的工作,通知还解决了何时执行这个工作的问题,即这个工作应该应用在某个方法被调用之前、之后、还是之前之后都调用。表 2-2 列举了通知的 5 种类型。

表 2-2　通知的 5 种类型

通　　知	说　　明
before(前置通知)	通知在目标方法调用之前执行
after(后置通知)	通知在目标方法调用之后执行
after-returning(返回通知)	通知在目标方法返回后执行
after-throwing(异常通知)	通知在目标方法抛出异常后执行
around(环绕通知)	通知将目标方法包裹起来，在目标方法调用之前和调用之后执行

4. 切点

切点(Pointcut)是匹配连接点的谓词。通知与切点表达式关联，并在与切点匹配的连接点上运行。如果说通知定义了切面的“做什么”和“什么时候做”的问题，那么切点就定义了“何处做”的问题。连接点是程序执行过程中能够应用通知的所有点；切点定义了通知被应用的具体位置(在哪些连接点应用通知)。

5. 引介

引介(Introduction)允许开发人员向现有的类添加新的方法或属性。例如，可以创建一个 Editable 通知类，该类记录了对象最后一次修改时的状态。这个类的对象可以被引入现有的类中，从而在无须修改现有类的情况下，让它们具有新的状态和行为。

6. 目标对象

目标对象(Target Object)指被插入切面的对象。

7. 代理

代理指将切面植入目标对象后由 AOP 框架创建的一个对象。

8. 织入

织入指将切面植入目标对象而形成代理对象的过程。

2.5 AspectJ AOP 注解开发

Spring 框架的 AOP 功能通常与 Spring IoC 一起使用。切面定义为普通的 Bean，这是 Spring 与其他 AOP 框架的一个关键区别。Spring AOP 仅支持执行公共(Public)非静态方法的调用作为连接点。如果需要将受保护的(Protected)或私有的(Private)的方法进行增强，就需要使用功能更全面的 AOP 框架来实现，其中使用最多的就是 AspectJ。AspectJ 是一个基于 Java 的全功能 AOP 框架，它并不是 Spring 的组成部分，是一个独立的 AOP 框架。

由于 AspectJ 支持通过 Spring 配置 AspectJ 切面，因此它是 Spring AOP 的完美补充。通常情况下，都是将 AspectJ 和 Spring 框架一起使用，以简化 AOP 操作。使用 AspectJ 需要在 Spring 项目中导入 Spring AOP 和 AspectJ 相关 JAR 包，在 pom. xml 文件中增加相关依赖。

```
1   <dependency>
2       <groupId>org.springframework</groupId>
3       <artifactId>spring-aspects</artifactId>
4       <version>${spring.version}</version>
5   </dependency>
```

```
6     <dependency>
7         <groupId>org.aspectj</groupId>
8         <artifactId>aspectjweaver</artifactId>
9         <version>1.9.7</version>
10    </dependency>
```

基于 AspectJ 实现 AOP 开发，分为基于 XML 的实现和基于注解的实现，本节讲授基于注解的 AOP 实现方法。

使用注解来创建切面是 AspectJ 5 引入的关键特性。通过 AspectJ 面向注解的模型可以非常方便地通过少量注解把任意类转变为切面。表 2-3 列举了 Spring 用于实现 AOP 的注解。

表 2-3 Spring AOP 注解

注 解	说 明
@After	配置后置(最终)通知
@AfterReturning	配置返回通知
@AfterThrowing	配置异常通知
@Around	配置环绕通知
@Aspect	配置切面
@Before	配置前置通知
@Pointcut	配置切点

从 Spring 5.2.7 开始，在同一@Aspect 类中定义的、需要在同一连接点上运行的通知方法根据其通知类型按以下顺序分配优先级：@Around、@Before、@After、@AfterReturning、@AfterThrowing。

下面通过一个案例来演示基于注解的 AOP 实现方法，实现步骤如下。

1. 创建接口

在 aspectj.annotation.dao 包中创建名为 CourseDao 的接口，代码如文件 2-7 所示。

【文件 2-7】 CourseDao.java

```
1    package aspectj.annotation.dao;
2    public interface CourseDao {
3        public void add();
4        public void delete();
5        public String list();
6        public int get();
7    }
```

2. 创建实现类

在 aspectj.annotation.dao.impl 包中创建 CourseDao 接口的实现类 CourseDaoImpl，代码如文件 2-8 所示。

【文件 2-8】 CourseDaoImpl.java

```
1    package aspectj.annotation.dao.impl;
2    //import 部分略
3    @Component
4    public class CourseDaoImpl implements CourseDao {
```

```
5        public void add() {
6          //配置前置通知
7          System.out.println("add course...");
8        }
9        public void delete() {
10         //配置环绕通知
11         System.out.println("remove course...");
12       }
13       public String list() {
14         //配置返回通知
15         System.out.println("course list: ...");
16         return "list";
17       }
18       public int get() {
19         //配置异常通知和最终通知
20         int a = 1/0;
21         return a;
22       }
23     }
```

3. 创建切面

在 aspectj. annotation. aspect 包中创建切面类 MyCourseAspect，并附加相关注解，代码如文件 2-9 所示。

【文件 2-9】 MyCourseAspect. java

```
1     package aspectj.annotation.aspect;
2     //import 部分略
3     @Component
4     @Aspect
5     public class MyCourseAspect {
6         @Before("execution( * aspectj.annotation.dao.impl.
7             CourseDaoImpl.add(..))")
8         public void before() {
9             System.out.println("前置通知......");
10        }
11
12        @After("execution( * aspectj.annotation.dao.impl.
13            CourseDaoImpl.get(..))")
14        public void after(){
15            System.out.println("后置(最终)通知......");
16        }
17
18        @Around("execution( * aspectj.annotation.dao.impl.
19            CourseDaoImpl.delete(..))")
20        public void around(ProceedingJoinPoint proceedingJoinPoint)
21            throws Throwable{
22            System.out.println("环绕通知 --- 前......");
23            proceedingJoinPoint.proceed();
24            System.out.println("环绕通知 --- 后......");
25        }
26
27        @AfterThrowing(pointcut = "execution( * aspectj.annotation.dao.
```

```
28              impl.CourseDaoImpl.get(..))", throwing = "exception")
29          public void afterThrow(Throwable exception){
30              System.out.println("异常通知...... 异常信息为: " +
31                  exception.getMessage());
32          }
33          @AfterReturning(pointcut = "execution( * aspectj.annotation.dao.
34              impl.CourseDaoImpl.list(..))", returning = "returnValue")
35          public void afterReturning(Object returnValue){
36              System.out.println("返回通知...... 方法返回值为: " + returnValue);
37          }
38      }
```

本例在形式上用注解定义了5个切点。如果要将不同类型的通知应用于同一个切点，也可以采用如下形式的定义：

```
1   public class MyCourseAspect {
2     @Pointcut("execution( * com.example.aop. * . * (..))")
3     public void myPointcut(){
4     }
5     @Before("myPointcut()")
6     public void beforeAdvice(){
7         ...
8     }
9     ...
10  }
```

此外，对于文件2-9中第27～32行的异常通知，由于通知方法afterThrow()要获取目标方法抛出的异常对象，在使用@AfterThrowing注解声明异常通知时，要附带throwing属性。该属性有两个作用：第一，限制抛出异常的类型为Throwable，这个类型与通知方法中参数类型一致；第二，将抛出的异常对象绑定为通知方法中的形参exception。

类似地，对于第33～37行的返回通知，由于通知方法afterReturning()要访问目标方法的返回值，在使用@AfterReturning注解声明返回通知时，要附带returning属性。该属性既限定了返回值的类型，又可以将返回值作为实参传递给通知方法继续处理。

4. 编写测试代码

为便于测试，需要编写一个配置类。这个类本身是一个标识，用来供@ComponentScan注解依附。@ComponentScan注解的作用是将相关的类配置为Spring中的Bean，并通知Spring容器扫描并创建这些Bean。配置类代码如文件2-10所示。

【文件2-10】 CourseConfig.java

```
1   package aspectj.annotation;
2   //import 部分略
3   @ComponentScan(basePackages = "aspectj.annotation")
4   @EnableAspectJAutoProxy
5   public class CourseConfig {
6   }
```

如文件2-10所示，@Configuration注解用来通知Spring容器，此类是配置类，是一个Bean的容器，相当于XML配置文件中的< beans >标记。@ComponentScan注解用来通知

Spring 容器进行包扫描的路径。Spring 会扫描 aspectj. annotation 包及其子包，并将实例化相关的类。AspectJ 不是基于代理的 AOP 框架，需要用@EnableAspectJAutoProxy 注解开启 AspectJ 的自动代理支持。

在 src/test/java 目录下编写测试代码，检查注解方式的方法增强。测试代码如文件 2-11 所示。

【文件 2-11】 TestAspectAnnotation. java

```
1   package com.example.spring.test;
2   //import 部分略
3   @RunWith(SpringJUnit4ClassRunner.class)
4   @ContextConfiguration(classes = AutoConfig.class)
5   public class TestAspectAnnotation {
6     @Autowired
7     private CourseDao courseDao;
8     @Test
9     public void testAnnotation(){
10        courseDao.add();
11        courseDao.delete();
12        courseDao.list();
13        courseDao.get();
14    }
15  }
```

以上步骤采用注解方式实现了 AspectJ AOP，利用通知对目标方法实现了增强。测试代码运行结果如图 2-5 所示。

```
前置通知……
add course...
环绕通知---前……
remove course...
环绕通知---后……
course list : ...
返回通知…… 方法返回值为: list
异常通知…… 异常信息为: / by zero
后置（最终）通知……
```

图 2-5 对目标方法增强后的效果

2.6 小结

Spring 的最核心理念是控制反转（IoC）和面向切面编程（AOP）。控制反转就是把对象的创建、销毁的权利交给 Spring，由 Spring 来管理对象的生命周期。这样，利用控制反转就可以实现组件间的松耦合。

面向切面编程就是采取横向抽取机制，把分散在各个方法中的相同功能的代码分离出来，形成切面，然后在编译阶段或运行时把切面应用到需要执行的地方。这样一方面分离了各方法中公共的辅助功能，实现了代码的进一步复用；另一方面，由于主要业务功能和辅助功能分离，使得开发人员可以聚精会神地关注主要业务功能的实现。

第 3 章

Spring Redis Template

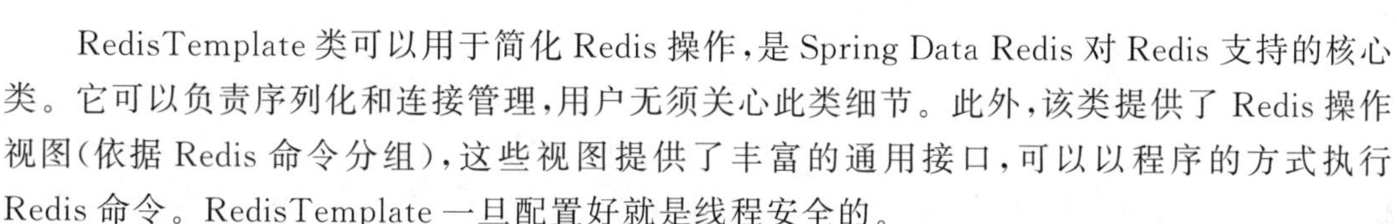

RedisTemplate 类可以用于简化 Redis 操作，是 Spring Data Redis 对 Redis 支持的核心类。它可以负责序列化和连接管理，用户无须关心此类细节。此外，该类提供了 Redis 操作视图(依据 Redis 命令分组)，这些视图提供了丰富的通用接口，可以以程序的方式执行 Redis 命令。RedisTemplate 一旦配置好就是线程安全的。

本章将围绕 RedisTemplate 类的连接建立、Redis 操作视图、键绑定、序列化等内容展开。

3.1 Java Redis 客户端

为了能够在应用程序中使用 Redis，通过 Redis 命令操作是不现实的。Redis 提供了适配不同编程语言的客户端实现通过编程操作 Redis。Java 有很多优秀的 Redis 客户端，如 Lettuce、Jedis、Redisson 和 JRedis 等。表 3-1 列举了常用的 Java Redis 客户端的特性。

表 3-1 常用的 Java Redis 客户端的特性

客户端	框架整合	介　　绍
Lettuce	Spring Data Redis	Lettuce 是基于 Netty 实现的，支持同步、异步和响应编程方式并且是线程安全的，支持 Redis 的哨兵模式、集群模式和流水线
Jedis	Spring Data Redis	以 Redis 命令作为方法名称，学习成本低，简单实用。Jedis 实例不是线程安全的，多线程环境下需要基于连接池来使用
Redisson	/	Redisson 是分布式 Redis 客户端，底层使用 Netty 框架，支持 Redis 的哨兵模式、主从模式和单节点模式

使用 Java 操作 Redis 最常用的是使用 Jedis 客户端。如果在项目中使用了 Jedis，但是后来决定弃用 Jedis 改用其他的 Redis 客户端就比较麻烦了。因为不同的 Java Redis 客户端是无法兼容的。Spring Data Redis 是 Spring Data 模块的一部分，专门用来支持在 Spring 管理的项目中对 Redis 的操作。Spring Data Redis 提供了 Redis 的 Java 客户端的抽象，在开发中可以忽略由于切换 Redis 客户端所带来的影响，而且它本身就属于 Spring 的一部分，比起单纯的使用 Jedis 更加稳定，管理起来更加自动化。

使用 Spring Data Redis 的首要任务之一是通过 Spring 容器连接到 Redis。为此，需要创建一个 Java 连接器。无论开发者选择哪个 Java Redis 客户端，只需要使用一组 Spring Data Redis API(在所有连接器中表现一致)。即使用 org. springframework. data. redis.

connection 包中 RedisConnection 和 RedisConnectionFactory 接口。这两个接口用于处理和获取 Redis 的活动连接。

RedisConnection 接口为应用程序与 Redis 通信提供了核心组件。因为它处理与 Redis 后端的通信。它还自动将底层连接库异常转换为与 Spring 一致的 DAO 异常层次结构，这样就可以在不更改任何代码的情况下切换连接器，因为操作语义保持不变。

RedisConnection 对象是通过 RedisConnectionFactory（工厂）创建的。此外，工厂充当 PersistenceExceptionTranslator 对象。这意味着一旦声明，PersistenceExceptionTranslator 对象就允许开发者进行透明的异常转换——例如，通过使用@Repository 注解和 AOP 进行异常转换。使用 RedisConnectionFactory 的最简单的方式就是通过 Spring 容器配置一个合适的连接器并将连接器注入给需要使用它的类。

Spring Data Redis 主要有以下特性：

(1) 提供了一个可以跨越多个客户端（如 Jedis、Lettuce）的底层抽象连接包。

(2) 针对数据的序列化和反序列化，提供了多种方案供开发者选择。

(3) 提供了一个 RedisTemplate 类，该类对 Redis 的各种操作、异常转换和序列化都实现了高层封装。

(4) 支持 Redis 的哨兵模式和集群模式。

3.2 创建 Redis 连接

针对不同的 Redis 客户端，使用程序连接 Redis 也有不同的方式。本节讲授如何利用 Lettuce、Jedis 客户端以及 Redis Template 类创建 Redis 连接。

3.2.1 Lettuce

Lettuce 是一个可扩展的线程安全 Redis 客户端，用于同步、异步和响应式使用。如果多个线程不使用阻塞和事务操作（如 BLPOP 和 MULTI/EXEC），则它们可能共享一个连接。Lettuce 是基于 Netty 构建的。Lettuce 支持 Redis 的高级功能，如哨兵（Sentinel）、集群（Cluster）、流水线（Pipelining）、自动重新连接和 Redis 数据模型。

要利用 Spring Data Redis 配置 Lettuce 连接器，首先在 Maven 项目中引入相关依赖，内容如下：

```
1   <!-- Spring 核心包 spring-context -->
2   <dependency>
3       <groupId>org.springframework</groupId>
4       <artifactId>spring-context</artifactId>
5       <version>5.3.18</version>
6   </dependency>
7   <dependency>
8       <groupId>org.springframework.data</groupId>
9       <artifactId>spring-data-redis</artifactId>
10      <version>2.7.1</version>
11  </dependency>
12  <!-- Lettuce -->
13  <dependency>
```

```
14        <groupId>io.lettuce</groupId>
15        <artifactId>lettuce-core</artifactId>
16        <version>6.1.8.RELEASE</version>
17    </dependency>
```

利用 Lettuce 配置 Redis 连接器有两种方案。第一，创建 Lettuce 连接工厂，通过连接工厂获取连接；第二，直接用 Lettuce 创建 Redis 连接。

1. Lettuce 连接工厂

可以创建一个名为 LettuceConfig 的类来配置 Lettuce 连接工厂，代码如文件 3-1 所示。

【文件 3-1】 LettuceConfig.java

```
1    @Configuration
2    public class LettuceConfig {
3      @Bean
4      public LettuceConnectionFactory redisConnectionFactory() {
5        return new LettuceConnectionFactory(
6             new RedisStandaloneConfiguration("127.0.0.1", 6379));
7      }
8    }
```

在创建了 LettuceConnectionFactory 实例后（第 5、6 行），可以调用该类的 getConnection() 方法获取 Redis 连接对象（org.springframework.data.redis.connection.RedisConnection）。同时，可以编写测试类查看获取的连接是否有效。测试代码如文件 3-2 所示。

【文件 3-2】 TestLettuce.java

```
1    @RunWith(SpringJUnit4ClassRunner.class)
2    @ContextConfiguration(classes = LettuceConfig.class)
3    public class TestLettuce {
4        @Autowired
5        private LettuceConnectionFactory factory;
6        @Test
7        public void testLettuceConnectionFactory(){
8            RedisConnection conn = factory.getConnection();
9            //向 Redis 发送 PING 命令,并断言得到 PONG
10           Assert.assertEquals("PONG",conn.ping());
11       }
12   }
```

如文件 3-2 所示，第 10 行调用 RedisConnectionCommands（RedisConnection 接口的父接口）接口的 ping()方法，用于向 Redis 发送 PING 命令（见 1.4.2 节 PING 命令部分）。

2. Lettuce 直接连接 Redis

创建一个名为 LettuceConnection 的类，代码如文件 3-3 所示。

【文件 3-3】 LettuceConnection.java

```
1    public class LettuceConnection {
2          public static void main(String[] args) {
3            //步骤 1: 连接信息
4            RedisURI redisURI = RedisURI.builder()
5               .withHost("127.0.0.1")
```

```
 6                .withPort(6379)
 7                //.withPassword(new char[]{'a', 'b', 'c'})
 8                .withTimeout(Duration.ofSeconds(10))
 9                .build();
10            //步骤2：创建Redis客户端
11            RedisClient client = RedisClient.create(redisURI);
12            //步骤3：建立连接
13            StatefulRedisConnection<String, String> connection =
14                    client.connect();
15            //向Redis发送操作命令,相关代码略
16            //关闭连接
17            connection.close();
18            client.shutdown();
19        }
20    }
```

3.2.2 Jedis

Jedis是利用Java操作Redis的工具,功能类似于JDBC。Jedis是Redis官方推荐的Java客户端开发包。Jedis支持Redis命令、事务和流水线。配置Jedis连接器,首先要在项目中添加Jedis的依赖,内容如下：

```
1    <!-- Jedis -->
2    <dependency>
3        <groupId>redis.clients</groupId>
4        <artifactId>jedis</artifactId>
5        <version>3.8.0</version>
6    </dependency>
```

利用Jedis作为Redis连接器有三种实现方案：Jedis连接工厂、Jedis直接连接和Jedis连接池。

1. Jedis连接工厂

创建一个JedisConfig类,代码如文件3-4所示。

【文件3-4】 JedisConfig.java

```
1    @Configuration
2    public class JedisConfig {
3        @Bean
4        public JedisConnectionFactory redisConnectionFactory() {
5            RedisStandaloneConfiguration config =
6                new RedisStandaloneConfiguration("127.0.0.1", 6379);
7            return new JedisConnectionFactory(config);
8        }
9    }
```

如文件3-4所示,在获取了JedisConnectionFactory实例后(第7行),可以调用该类的getConnection()方法获取Redis连接。

2. Jedis直接连接

所谓直接连接是指Jedis在每次发送Redis操作命令前都会新建TCP连接,使用后再

断开连接，如图 3-1 所示。对于频繁访问 Redis 的场景这种方案，显然不是高效的使用方式。

利用 Jedis 直接创建 Redis 连接非常简单，只需创建 Jedis 的实例并指定相关参数即可。下述代码描述了图 3-1 的完整过程。

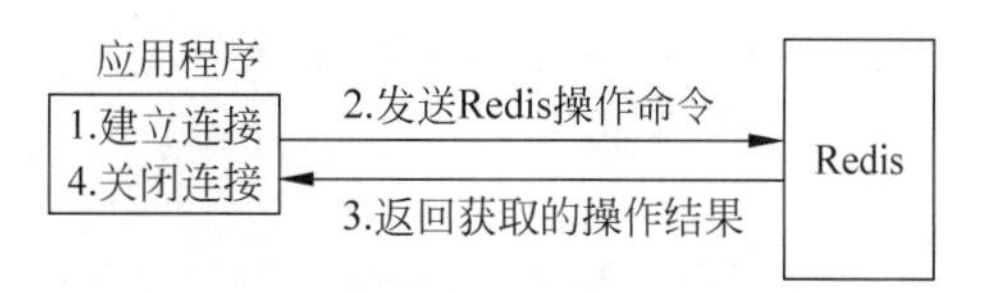

图 3-1 Jedis 直接连接 Redis

```
1    //1.建立连接
2    Jedis jedis = new Jedis("localhost",6379);
3    //2.发送 Redis 操作命令
4    jedis.set("username","zhangsan");
5    //3.返回获取的操作结果
6    String username = jedis.get("username");
7    //4.关闭连接
8    jedis.close();
```

3. Jedis 连接池

在生产环境中，从提升性能的角度考虑，一般使用连接池方式管理 Jedis 连接。具体做法是将所有的 Jedis 连接对象预先存放在连接池中，每次连接 Redis 时，只需要从连接池中借用活动连接，用后再将连接归还给连接池，如图 3-2 所示。

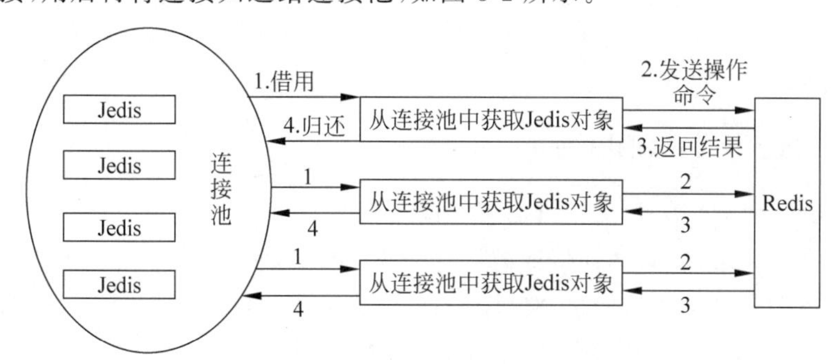

图 3-2 Jedis 连接池

客户端连接 Redis 使用的协议是 TCP。直接连接的方式每次都需要建立 TCP 连接，而连接池方式是可以预先创建好 Redis 连接，所以每次只需从连接池中借用即可。而借用和归还操作都是在本地进行的，只有少量的并发同步开销，这个开销远远小于新建 TCP 连接的开销。另外，直接连接方式无法限制 Jedis 对象的个数，在极端情况下会造成连接泄露，而连接池方式可以有效地保护和控制资源的使用。表 3-2 给出了两种方式各自的优势和劣势。

表 3-2 Jedis 直接连接方式和连接池方式对比

	优 点	缺 点
直接连接	简单方便，适用于少量长期连接的场景	(1) 存在每次新建、关闭 TCP 连接的开销； (2) 资源无法控制，极端情况会出现连接泄露； (3) Jedis 对象线程不安全
连接池	(1) 无须每次连接都生成 Jedis 对象，降低开销； (2) 保护和控制资源的使用	使用相对麻烦，尤其在资源管理上需要很多参数来保证，一旦规划不合理就会出现问题

创建 Redis 连接池代码如文件 3-5 所示。

【文件 3-5】 JedisConnectionPool. java

```
1    public class JedisConnectionPool {
2        private static final JedisPool jedisPool;
3        static{
4          GenericObjectPoolConfig jedisPoolConfig =
5                 new GenericObjectPoolConfig();
6          //连接池中的最大连接数
7          jedisPoolConfig.setMaxTotal(8);
8          //连接池中的最大空闲连接数
9          jedisPoolConfig.setMaxIdle(8);
10        //连接池中的最少空闲连接数
11        jedisPoolConfig.setMinIdle(0);
12          //当连接资源耗尽后,调用者的最大等待时间,单位：毫秒.－1 表示永不超时
13          jedisPoolConfig.setMaxWait(Duration.ofMillis(200));
14          jedisPool =
15               new JedisPool(jedisPoolConfig,"127.0.0.1",6379,1000);
16        }
17        //从连接池中借用连接
18        public static Jedis getJedis(){
19            return jedisPool.getResource();
20        }
21   }
```

3.2.3 RedisTemplate

目前,Jedis 客户端在编程实施方面存在以下一些不足。

（1）连接管理无法自动化,连接池的设计缺少必要的容器支持。

（2）数据操作需要关注序列化和反序列化,因为 Jedis 的客户端 API 接受的数据类型为 String 和 Byte,对结构化数据(JSON、XML、POJO 等)操作需要额外的支持。

（3）事务操作为硬编码。

（4）对于发布-订阅功能缺乏必要的设计模式支持,对于开发者而言需要关注的内容太多。

RedisTemplate 类(org. springframework. data. redis. core. RedisTemplate)是 Spring Data Redis 中对 Jedis API 的高度封装。Spring Data Redis 相对于 Jedis 来说可以方便地更换 Redis 的 Java 客户端。与 Jedis 相比,Spring Data Redis 进行了如下改进。

（1）连接池自动管理,提供了一个高度封装的 RedisTemplate 类。

（2）针对 Jedis 客户端大量的 API 进行了归类封装。遵循 Redis 命令参考(见 Redis 官方网站)中的分组,RedisTemplate 提供包含丰富的通用子接口的操作视图。这些视图可用于针对特定类型或特定键(通过键绑定接口)进行操作。

（3）由容器封装并控制事务操作。

（4）针对数据的序列化和反序列化提供了多种可选择的序列化器(RedisSerializer)。

（5）基于设计模式和 JMS(Java Message Service,Java 消息服务)开发思路,将发布-订阅的编程接口进行了封装,使开发更加便捷。

（6）RedisTemplate 是线程安全的。

本书的后续案例将使用 RedisTemplate 来实现编程执行 Redis 操作。要获取 RedisTemplate 实例，首先要创建连接工厂对象，再借助连接工厂构建 RedisTemplate。如果操作的值是 String 类型，也可以使用 RedisTemplate 类的子类 StringRedisTemplate。由于程序中要频繁使用 RedisTemplate 对象，可以将其设置为 Spring 管理的 Bean，然后由 Spring 将该对象注入需要的地方。代码如文件 3-6 所示。

【文件 3-6】 RedisTemplateConfig. java

```
1   package com.example.redis.template;
2   //import 部分略
3   @Configuration
4   public class RedisTemplateConfig {
5       @Bean
6       publicRedisConnectionFactory redisConnectionFactory() {
7           RedisStandaloneConfiguration rsc =
8               new RedisStandaloneConfiguration();
9           rsc.setHostName("127.0.0.1");
10          rsc.setDatabase(0);
11          rsc.setPort(6379);
12          return new JedisConnectionFactory(rsc);
13      }
14      @Bean
15      public StringRedisTemplate redisTemplate(@Autowired
16      RedisConnectionFactory rcf){
17          StringRedisTemplate template = new StringRedisTemplate();
18          template.setConnectionFactory(rcf);
19          return template;
20      }
21  }
```

从 2.0 版本开始，Spring Data Redis 已经不推荐直接显式设置连接的信息了，一方面为了使配置信息与建立连接工厂解耦，另一方面抽象出 Standalone、Sentinel 和 RedisCluster 三种模式的环境配置类与一个统一的 Jedis 客户端连接配置类（用于配置连接池和 SSL（Secure Socket Layer，安全套接字层）连接），这样可以更加灵活、方便地根据实际业务场景需要来配置连接信息。文件 3-6 以 Standalone 方式为例，展示了在不使用连接池的情况下，如何实例化 RedisTemplate。首先创建 RedisStandaloneConfiguration 实例并设置参数（第 7～11 行），然后根据该配置实例来初始化 Jedis 连接工厂（第 12 行）。

文件 3-6 的配置使用的是直接连接 Redis 的方式，即每次需要时都会创建新的连接。当并发量剧增时，这会带来性能上的开销，同时由于没有对连接数进行限制，可能使服务器崩溃导致无法响应。所以一般会建立连接池，事先初始化一组连接，供需要 Redis 连接的线程取用。采用连接池方式需要更改文件 3-6 的第 5～13 行，具体如下：

```
1   @Bean
2   public RedisConnectionFactory redisConnectionFactory() {
3      JedisPoolConfig jpc = new JedisPoolConfig();
4      jpc.setMaxTotal(8);
5      jpc.setMaxIdle(8);
6      jpc.setMinIdle(0);
```

```
7         jpc.setMaxWait(Duration.ofMillis(200));
8         //Redis 连接配置
9         RedisStandaloneConfiguration redisStandaloneConfiguration =
10            new RedisStandaloneConfiguration();
11        //设置 Redis 服务器的 IP
12        redisStandaloneConfiguration.setHostName("127.0.0.1");
13        //设置 Redis 服务器的端口号
14        redisStandaloneConfiguration.setPort(6379);
15        //连接的数据库
16        redisStandaloneConfiguration.setDatabase(0);
17        //JedisConnectionFactory 配置 jedisPoolConfig
18        JedisClientConfiguration.JedisClientConfigurationBuilder
19          jedisClientConfiguration = JedisClientConfiguration.builder();
20        //指定连接池
21        jedisClientConfiguration.usePooling().poolConfig(jpc);
22        //创建工厂对象
23        RedisConnectionFactory factory = new JedisConnectionFactory(
24          redisStandaloneConfiguration, jedisClientConfiguration.build());
25        return factory;
26    }
```

在默认情况下，Redis 服务器在启动时会创建 16 个数据库，编号从 0 到 15。不同的应用可以连接到不同的数据库上。上述代码的第 16 行通过 setDatabase()方法选择编号为 0 的数据库。此外，Spring Data Redis 提供的采用连接池创建 RedisTemplate 对象的方式并不优雅。如果要采用连接池创建 RedisTemplate 对象，推荐使用 Spring Boot。

RedisTemplate 默认使用 Java 的序列化程序。通过 RedisTemplate 写入或读取的任何对象都是通过 Java 序列化和反序列化的。可以通过 org.springframework.data.redis.serializer 包中提供的接口更改默认的序列化机制的设置，内容可见 3.12 节。

3.3 Spring 操作 Redis 字符串

在创建了 RedisTemplate（或 StringRedisTemplate）对象后，可以编程执行 Redis 操作了。Redis 可以存取多种不同类型的数据，其中有 5 种基础数据类型：字符串、列表、哈希、集合和有序集合。此外，还有流、地理空间数据、位图等。RedisTemplate 对基础数据类型的大部分操作都是借助表 3-3 中的方法和子接口完成的。

表 3-3 RedisTemplate 操作基础数据类型的主要方法和子接口

方法	子接口	描述
opsForValue()	ValueOperations<K,V>	操作字符串类型的条目
opsForList()	ListOperations<K,V>	操作列表类型的条目
opsForSet()	SetOperations<K,V>	操作集合类型的条目
opsForHash()	HashOperations<K,V>	操作哈希类型的条目
opsForZSet()	ZSetOperations<K,V>	操作有序集合类型的条目
opsForStream()	StreamOperations<K,HK,HM>	执行流操作命令的接口
opsForHyperLogLog()	HyperLogLogOperations<K,V>	操作超级日志
opsForGeo()	GeoOperations<K,M>	操作地理空间数据类型的条目

要操作 Redis 字符串，需要调用 RedisTemplate 类的 opsForValue()方法创建 ValueOperations 子接口对象，再调用 ValueOperations 子接口的相关方法。本节介绍 ValueOperations 子接口(org.springframework.data.redis.core.ValueOperations<K,V>)中的主要方法的使用。

(1) 方法原型：void set(K key,V value)；功能：设置键 key 的值 value；对应 Redis 命令：SET。

(2) 方法原型：@Nullable V get(Object key)；功能：返回键 key 的值，当键的值不存在或在流水线(或事务)中使用该方法时，返回 null；对应 Redis 命令：GET。

【例 3-1】 利用键 user 保存值 hello,redis。可以通过 ValueOperations 子接口提供的 set(String key,String value)方法实现，代码如文件 3-7 所示。

【文件 3-7】 MyRedisStringTest.java

```
1    @RunWith(SpringJUnit4ClassRunner.class)
2    @ContextConfiguration(classes= RedisTemplateConfig.class)
3    public class MyRedisStringTest {
4      @Autowired
5      private StringRedisTemplate redis;
6      @Test
7      public void testEx1() {
8          redis.opsForValue().set("user","hello,redis");
9          String str = redis.opsForValue().get("user");
10         assertEquals("hello,redis",str);
11     }
12   }
```

如文件 3-7 所示，本节的案例由于操作的值都是 String 类型的，因此采用了 RedisTemplate 类的子类 StringRedisTemplate。StringRedisTemplate 对象在文件 3-6 的第 14～20 行完成实例化，并注入测试类中(第 4、5 行)。表 3-3 中的子接口对象可以通过 RedisTemplate 或 StringRedisTemplate 创建。即通过调用 RedisTemplate 的 opsForValue()方法获得子接口引用(第 8 行)，再调用子接口中对应的方法完成字符串操作。第 8 行调用 set()方法向 Redis 中以 user 为键，写入值 hello,redis。第 9 行通过键 user 取出对应的值。

(3) 方法原型：void set(K key,V value,long timeout,TimeUnit unit)；功能：设置键 key 的值 value 和过期超时时长 timeout；对应 Redis 命令：SETEX。

【例 3-2】 以 name 为键将值 Tom 写入 Redis，并设置写入键值的有效时长为 10 秒。测试代码如文件 3-8 所示。

【文件 3-8】 例 3-2 测试代码

```
1    @Test
2    public void testEx2() throws Interrupted Exception {
3       redis.opsForValue().set("name","Tom",10,TimeUnit.SECONDS);
4       Thread.sleep(11000);
5       String str = redis.opsForValue().get("name");
6       assertEquals("Tom",str);
7    }
```

在设置 name 键值后，等待 10 秒再去获取 name 键的值，get()方法将返回 null，测试代码运行结果如图 3-3 所示。

```
java.lang.AssertionError:
Expected :Tom
Actual   :null
```

图 3-3 例 3-2 测试代码运行结果

（4）方法原型：void set(K key,V value,long offset)；功能：对于键 key 所存储的字符串；从指定偏移量 offset 开始用给定值 value 替代对应 Redis 命令：SETRANGE。

【例 3-3】 向 Redis 中以 key 为键存入字符串 hello，world，随后将该字符串替换为 hello，redis。测试代码如文件 3-9 所示。

【文件 3-9】 例 3-3 测试代码

```
1    @Test
2    public void testEx3(){
3        redis.opsForValue().set("key","hello,world");
4        redis.opsForValue().set("key","redis",6);
5        String str = redis.opsForValue().get("key");
6        assertEquals("hello,redis",str);
7    }
```

（5）方法原型：@Nullable Boolean setIfAbsent(K key,V value)；功能：如果缺少键 key，则设置以键 key 保存字符串值 value；对应的 Redis 命令：SETNX。

【例 3-4】 以 abs 为键，调用 setIfAbsent()方法保存字符串 absent，保存后将值改为 present。测试代码如文件 3-10 所示。

【文件 3-10】 例 3-4 测试代码

```
1    @Test
2    public void testEx4(){
3        assertTrue(redis.opsForValue().setIfAbsent("abs","absent"));
4        assertFalse(redis.opsForValue().setIfAbsent("abs","present"));
5    }
```

如文件 3-10 所示，如果 Redis 中不存在键 abs，则第 3 行代码执行成功。第 4 行准备将键 abs 的值修改为 present。由于键 abs 已存在，因此此次更改失败。

（6）方法原型：void multiSet(Map <? extends K，? extends V > map)；功能：使用元组中提供的键值对将多个键设置为多个值；对应的 Redis 命令：MSET。

（7）方法原型：@Nullable List < V > multiGet(Collection < K > keys)；功能：返回所有给定键的值，值按键的请求顺序返回；对应的 Redis 命令：MGET。

【例 3-5】 将字符串 aaa、bbb 和 ccc 一次性存入 Redis，这三个字符串对应的键分别为 multi1、multi2 和 multi3，再通过各自的键将它们取出。测试代码如文件 3-11 所示。

【文件 3-11】 例 3-5 测试代码

```
1    @Test
2    public void testEx5(){
3        Map < String, String > map = new HashMap < String, String >();
4        map.put("multi1","aaa");
5        map.put("multi2","bbb");
6        map.put("multi3","ccc");
7        redis.opsForValue().multiSet(map);
8
```

```
 9         List<String> list = new ArrayList<String>();
10         list.add("multi1");
11         list.add("multi2");
12         list.add("multi3");
13         List<String> values = redis.opsForValue().multiGet(list);
14         values.forEach(System.out::println);
15     }
```

（8）方法原型：@Nullable V getAndSet(K key,V value)；功能：设置键 key 的值并返回其旧值；对应的 Redis 命令：GETSET。

【例 3-6】 应用 getAndSet()方法。测试代码如文件 3-12 所示。

【文件 3-12】 例 3-6 测试代码

```
1   @Test
2   public void testEx6(){
3      redis.opsForValue().set("getset","test-11");
4      String str = redis.opsForValue().getAndSet("getset","test-22");
5      assertEquals("test-11",str);
6   }
```

（9）方法原型：@Nullable Long increment(K key,long delta)；功能：将存储在键 key 下的字符串的整数值按增量 delta 递增，如果 key 指定的值不存在，那么 key 的值会先被初始化为 0，然后再执行递增；对应的 Redis 命令：INCRBY。

【例 3-7】 点赞是社交网络中最常用的功能。本例模拟社交网络中对作品的点赞功能，并输出当前作品获赞的数量。测试代码如文件 3-13 所示。

【文件 3-13】 例 3-7 测试代码

```
 1   @Test
 2   public void testEx7(){
 3       boolean laud_flag = true;
 4       Long l = 0L;
 5       if(laud_flag)
 6           l = redis.opsForValue().increment("articleId",1);
 7       else
 8           l = redis.opsForValue().increment("articleId",-1);
 9       System.out.println(l);
10   }
```

将点赞数量存入 Redis，以作品 Id(标识符属性，字符串类型)为键。如果用户点赞，则变量 laud_flag 取值为 true，对应的点赞数增加 1；反之，用户取消点赞，变量 laud_flag 取值为 false，对应的点赞数减 1。

（10）方法原型：@Nullable Boolean setBit(K key,long offset,boolean value)；功能：对键 key 所存储的字符串值，设置或清除指定偏移量上的位；对应的 Redis 命令：SETBIT。

（11）方法原型：@Nullable Boolean getBit(K key,long offset)；功能：获取键对应值的 ASCII 码在 offsest 处的值；对应的 Redis 命令：GETBIT。

【例 3-8】 利用键 bit 存入字符串 a，并利用位运算将该字符串改为 b。测试代码如文件 3-14 所示。

【文件 3-14】 例 3-8 测试代码

```
1    @Test
2    public void testEx8(){
3        redis.opsForValue().set("bit","a");
4        assertTrue(redis.opsForValue().getBit("bit",7));
5        redis.opsForValue().setBit("bit",6,true);
6        redis.opsForValue().setBit("bit",7,false);
7        assertEquals("b",redis.opsForValue().get("bit"));
8    }
```

上述测试代码第 3 行以 bit 为键存入字符串 a。字符'a'的 ASCII 码是 97，其二进制形式为 01100001。这样，第 4 行获取第 7 位的 ASCII 码值，得到二进制 1。1 代表 true，0 代表 false。因此，第 4 行结果应为 true。第 5、6 行分别将字符'a'的 ASCII 码的第 6、7 位设置为 1 和 0，这样键 bit 对应的 ASCII 码被改为 01100010，即字符'b'的 ASCII 码。

目前的软件系统(包括电商和社交网络)经常有这样的需求：根据用户提供的手机号码发送验证码，实现登录。下面的案例模拟实现手机验证码的发送功能。

【例 3-9】 利用 Redis 实现模拟手机验证码登录之验证码发送功能。对于验证码的发送，通常有如下要求：

(1) 发送的手机验证码几分钟内(本例设定为 1 分钟)有效。

(2) 每天向每个手机号码发送的验证码的次数有限(本例设定为 24 小时内最多发送 3 次)。

为实现上述两项要求，可以利用 Redis 保存两个值，并设置它们的生存时间：一个用于保存发送给用户的验证码，生存时长为 1 分钟；另一个用来保存给用户发送验证码的次数，生存时间为 24 小时。测试代码如文件 3-15 所示。

【文件 3-15】 ShortMessageSender.java

```
1    public class ShortMessageSender {
2        public String sendCode(StringRedisTemplate template,String phone){
3            String codeKey = phone + "_CODE";
4            String countKey = phone + "_COUNT";
5            Integer count = 0;
6            try {
7                count = Integer.parseInt(
8                    template.opsForValue().get(countKey));
9            } catch(NumberFormatException nfe) {
10               count = 0;
11           }
12           if(count > 2) {
13               System.out.println("24 小时内发送次数已达 3 次,24 小时后重试");
14               return "retry";
15           }
16           Boolean hasCodeKey = template.hasKey(codeKey);
17           if(hasCodeKey){
18               Long codeTTL = template.getExpire(codeKey);
19               System.out.println("验证码 1 分钟内有效,请在" + codeTTL +
20                   "秒之后再次发送");
21               return "tip";
```

```
22            }
23            String code = RandomUtil.randomNumbers(6);
24            System.out.println("CODE IS : " + code);
25            template.opsForValue().set(codeKey,code,60, TimeUnit.SECONDS);
26            long timeout = 24 * 60 * 60;
27            if(count!= 0)
28                timeout = template.getExpire(countKey);
29             template.opsForValue().set(countKey,String.valueOf(count + 1),
30                 timeout,TimeUnit.SECONDS);
31             System.out.println("验证码发送成功");
32             return "success";
33        }
34    }
```

如文件 3-15 所示，对于用户提交的手机号码，以参数 phone 传入 sendCode()方法，并省去了对手机号码合法性的校验。第 3、4 行分别以手机号码附带后缀的形式定义了两个键，这两个键对应的值分别为验证码和发送验证码的次数。第 7、8 行试图从 Redis 中获取发送验证码的次数，当 Redis 中不存在键 countKey 时，将记录验证码发送次数的变量 count 计为 0(第 9～11 行)。第 23 行采用 RandomUtil 工具随机生成包含 6 位数字的验证码。要使用 RandomUtil 工具，需要在 pom.xml 文件中添加相关依赖：

```
1    <!-- hutool -->
2    <dependency>
3        <groupId>cn.hutool</groupId>
4        <artifactId>hutool-all</artifactId>
5        <version>5.8.20</version>
6    </dependency>
```

第 25 行将验证码保存到 Redis，以便用户提交验证码后进行比对。同时，设置了验证码在 Redis 中的保存时间为 60 秒。第 29 行用同样的方法将验证码的发送次数保存到 Redis，同时设置该值在 Redis 中的保存时间为 24 小时。

随后，模拟向号码为 15612345678 的手机发送验证码，测试代码如文件 3-16 所示。

【文件 3-16】 TestSMSVerification.java

```
1     @RunWith(SpringJUnit4ClassRunner.class)
2     @ContextConfiguration(classes = RedisTemplateConfig.class)
3     public class TestSMSVerification {
4         @Autowired
5         private StringRedisTemplate template;
6
7         @Test
8         public void testSender(){
9             ShortMessageSender sender = new ShortMessageSender();
10            sender.sendCode(template,"15612345678");
11        }
12    }
```

测试代码运行结果如图 3-4 所示，再次运行测试代码，结果如图 3-5 所示。

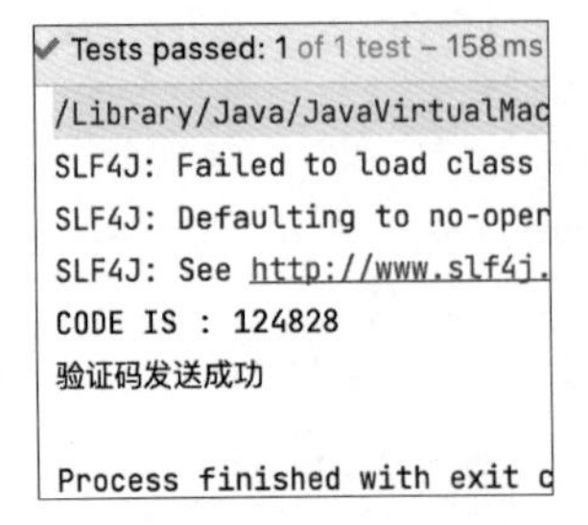

图 3-4 文件 3-16 测试代码运行结果

```
Tests passed: 1 of 1 test – 159 ms
/Library/Java/JavaVirtualMachines/
SLF4J: Failed to load class "org.s
SLF4J: Defaulting to no-operation
SLF4J: See http://www.slf4j.org/co
验证码1分钟内有效，请在33秒之后再次发送

Process finished with exit code 0
```

图 3-5 再次运行测试代码的结果

3.4 Spring 操作 Redis 列表

在 Redis 中，列表类型是按照元素插入的顺序排序的字符串列表。可以向列表的头部（左边）或尾部（右边）添加、删除元素。操作 Redis 列表的方法与操作 Redis 字符串的方法类似，要调用 RedisTemplate 类的 opsForList()方法创建 ListOperations 子接口对象，再调用 ListOperations 子接口的相关方法。本节介绍 ListOperations 子接口（org. springframework. data. redis. core. ListOperations < K，V >）中的主要方法的使用。

（1）方法原型：@Nullable Long leftPush（K key，V value）；功能：将一个或多个值 value 插入列表 key 的表头，特点是先进后出，可以作为栈使用；返回值：执行插入操作后的列表长度；对应的 Redis 命令：LPUSH。

（2）方法原型：@Nullable Long rightPush（K key，V value）；功能：将一个或多个值 value 插入列表 key 的表尾，特点是先进先出，可以作为队列使用；返回值：执行插入操作后的列表长度；对应的 Redis 命令：RPUSH。

（3）方法原型：@Nullable List < V > range（K key，long start，long end）；功能：获取指定范围内的元素列表，参数 start 和 end 分别代表开始索引和结束索引。索引从左到右分别为 0 到 $N-1$，从右到左为 -1 到 $-N$。并且，end 参数包含了自身。对应的 Redis 命令：LRANGE。

【例 3-10】 将数字 1～10 以 strs 为键保存到 Redis 中，并倒序输出。测试代码如文件 3-17 所示。

【文件 3-17】 例 3-10 测试代码

```
1    @Test
2    public void testEx9() {
3       for(int i = 1;i < 11;i++)
4          redis.opsForList().leftPush("strs",String.valueOf(i));
5       redis.opsForList().range("strs", 0, -1)
6          .forEach(e -> System.out.print(e + " "));
7    }
```

```
10 9 8 7 6 5 4 3 2 1
```

图 3-6 例 3-10 测试代码运行结果

执行测试代码，运行结果如图 3-6 所示。从控制台输出可见 leftPush()方法实现了栈操作。读者可自行修改代码，实现正序输出。

（4）方法原型：@Nullable Long rightPushAll(K key,Collection < V > values)；功能：将一组值插入列表 key 的尾部；返回值：执行插入后的列表长度；对应的 Redis 命令：RPUSH。

（5）方法原型：@Nullable Long leftPushAll(K key,Collection < V > values)；功能：将一组值插入列表 key 的头部；返回值：执行插入后的列表长度；对应的 Redis 命令：LPUSH。

【例 3-11】 将数字 1～10 以 strs 为键批量保存到 Redis 中，并正序输出。测试代码如文件 3-18 所示。

【文件 3-18】 例 3-11 测试代码

```
1   @Test
2   public void testEx10() {
3      List < String > strs = new ArrayList < String >();
4      for(int i = 1;i < 11;i++)
5         strs.add(String.valueOf(i));
6      redis.opsForList().rightPushAll("strs",strs);
7      redis.opsForList().range("strs", 0, - 1)
8         .forEach(System.out::println);
9   }
```

（6）方法原型：void trim(K key,long start,long end)；功能：修剪列表，使其保留 start 到 end 之间的值；对应的 Redis 命令：LTRIM。

【例 3-12】 修剪例 3-10 中的 strs 列表，保留 6～10。测试代码如文件 3-19 所示。

【文件 3-19】 例 3-12 测试代码

```
1   @Test
2   public void testEx11() {
3      System.out.println(redis.opsForList().range("strs",0, - 1));
4      redis.opsForList().trim("strs",5, - 1);
5      System.out.println(redis.opsForList().range("strs",0, - 1));
6   }
```

测试代码运行结果如图 3-7 所示。

```
[1, 2, 3, 4, 5, 6, 7, 8, 9, 10]
[6, 7, 8, 9, 10]
```

图 3-7 例 3-12 测试代码运行结果

（7）方法原型：@Nullable V leftPop(K key)；功能：移除并返回列表 key 中的第一个元素；对应的 Redis 命令：LPOP。

（8）方法原型：@Nullable V rightPop(K key)；功能：移除并返回列表 key 中的最后一个元素；对应的 Redis 命令：RPOP。

【例 3-13】 利用 Redis List 数据类型实现栈和队列。测试代码如文件 3-20 所示。

【文件 3-20】 例 3-13 测试代码

```
1   @Test
2   public void testEx12(){
3      String s = "abcde";
4      for(int i = 0;i < s.length();i++)
5         redis.opsForList().rightPush("letters",
```

```
6               String.valueOf(s.charAt(i)));
7       System.out.println("the current list:"
8           + redis.opsForList().range("letters",0, - 1));
9       System.out.println("As a stack");
10      for(int i = 0;i < 5;i++)
11          //Stack
12          System.out.println(redis.opsForList().rightPop("letters"));
13          //Queue
14          //System.out.println(redis.opsForList().leftPop("letters"));
15  }
```

如文件 3-20 所示，要实现栈，只要保证数据在列表的同一端执行插入和删除操作。为使字母入栈顺序与字母表序一致，采用了 rightPush()方法（第 5、6 行）在右侧执行插入操作。同时，在右侧执行删除操作（第 12 行）。测试代码运行结果如图 3-8 所示。

```
the current list:[a, b, c, d, e]
As a stack :
e d c b a
```

图 3-8 例 3-13 测试代码运行结果

（9）方法原型：@Nullable V move(K sourceKey, RedisListCommands.Direction from, K destinationKey, RedisListCommands.Direction to)；功能：自动返回并删除存储在列表 sourceKey 中的第一个或最后一个元素（表头或表尾，取决于 from 参数），并将该元素推送到列表 destinationKey 的第一个或最后一个元素（表头或表尾，取决于 to 参数），该方法需要 Redis 6.2.0 及以上版本支持；对应的 Redis 命令：LMOVE。

【例 3-14】 向 list1 中添加三个值，分别为 one、two 和 three，随后将 three 和 one 从 list1 中移除，并分别移入 list2 的表头和表尾。测试代码如文件 3-21 所示。

【文件 3-21】 例 3-14 测试代码

```
1   @Test
2   public void testEx13() {
3       String[] s = {"one","two","three"};
4       redis.opsForList().rightPushAll("list1",s);
5       redis.opsForList().move("list1", RedisListCommands.Direction.RIGHT
6           ,"list2",RedisListCommands.Direction.LEFT);
7       redis.opsForList().move("list1",RedisListCommands.Direction.LEFT
8           ,"list2",RedisListCommands.Direction.RIGHT);
9       System.out.println(redis.opsForList().range("list1",0, - 1));
10      System.out.println(redis.opsForList().range("list2",0, - 1));
11  }
```

上述代码的第 5、6 行调用 move()方法，首先将 list1 中的尾部元素 three 删除，随后将其移入 list2 的头部。类似地，第 7、8 行调用 move()方法将 list1 中的头部元素 one 删除，随后将其移入 list2 的尾部。运行此测试代码，结果如图 3-9 所示。

```
[two]
[three, one]
```

图 3-9 例 3-14 测试代码运行结果

Redis 通常用作消息传递服务器，用于处理后台作业或其他类型的消息传递任务。一种简单的消息队列形式通常是将值推送到生产者端的列表中，等待消费者端使用 RPOP（使用轮询）命令使用该值。然而，这种消息队列并不可靠，因为消息可能会丢失。例如，网络存在传输问题的情况，或者如果消费者在收到消息后不久崩溃，但消息尚未被处理。

LMOVE(或 BLMOVE 用于阻塞变体)命令提供了一种避免此问题的方法：消费者获取消息,同时将其推送到待处理列表中。一旦消息被处理,消费者将使用 LREM 命令从处理列表中删除消息。还可以使用另一个客户端监视处理列表中的项目是否保留太长时间,并在需要时将这些超时的项目再次推送到消息队列中。

此外,利用 move()方法还可以实现访问 *N* 个元素列表中的每个元素,而无须使用 LRANGE 命令将完整列表从服务器传输到客户端。

【例 3-15】 利用 move()方法遍历列表 cirList。测试代码如文件 3-22 所示。

【文件 3-22】 例 3-15 测试代码

```
1    @Test
2    public void testEx14() {
3       String[] s = {"one","two","three"};
4       redis.opsForList().rightPushAll("cirList",s);
5       Long size = redis.opsForList().size("cirList");
6       for(int i = 0;i < size;i++)
7          System.out.println(redis.opsForList()
8                .move("cirList",RedisListCommands.Direction.RIGHT
9                ,"cirList",RedisListCommands.Direction.LEFT));
10   }
```

其中,第 5 行调用 size()方法获取列表 cirList 中元素的个数。其方法原型为@Nullable Long size(K key)。运行此测试代码,结果如图 3-10 所示。

```
three
two
one
```

图 3-10 例 3-15 测试代码运行结果

(10) 方法原型：void set(K key,long index,V value)；功能：在索引 index 处设置列表元素的值；对应的 Redis 命令：LSET。

(11) 方法原型：@Nullable V index(K key,long index)；功能：获取列表 key 中索引为 index 的元素的值。对应的 Redis 命令：LINDEX。

【例 3-16】 将例 3-12 中结果列表的最后一个元素的值改为 100,并获取修改后的元素值。测试代码如文件 3-23 所示。

【文件 3-23】 例 3-16 测试代码

```
1    @Test
2    public void testEx15() {
3       redis.opsForList().set("strs",4,"100");
4       String n = redis.opsForList().index("strs",4);
5       assertEquals("100",n);
6    }
```

本章将 Redis 操作中的值都设定为 String 类型。因此,在上述代码的第 3 行,更改的新值也以字符串的形式表示。

(12) 方法原型：@Nullable Long remove(K key,long count,Object value)；功能：根据参数 count 的值,移除列表中与参数 value 相等的元素。count 的值可以是以下几种：

① count>0:从表头开始向表尾搜索,移除值与 value 相等的元素,数量为 count。

② count<0:从表尾开始向表头搜索,移除值与 value 相等的元素,数量为 count 的绝对值。

③ count=0：移除表中所有值与 value 相等的值。

④ 返回值：被移除的元素的数量。因为不存在的 key 被视作空表，所以当 key 不存在时，该方法返回 0。对应的 Redis 命令：LREM。

【例 3-17】 移除列表中的重复值。测试代码如文件 3-24 所示。

【文件 3-24】 例 3-17 测试代码

```
1    @Test
2    public void testEx16() {
3        String[] s = {"hello","hello","foo","hello"};
4        redis.opsForList().rightPushAll("lrm",s);
5        System.out.println(redis.opsForList().range("lrm",0, - 1));
6        redis.opsForList().remove("lrm", - 2,"hello");
7        System.out.println(redis.opsForList().range("lrm",0, - 1));
8    }
```

```
[hello, hello, foo, hello]
[hello, foo]
```

图 3-11 例 3-17 测试代码运行结果

运行此测试代码，结果如图 3-11 所示。

社交网络（或电商系统）中经常有这样的需求，用户可以查看浏览内容的历史记录。如果只是要求保留用户的浏览记录，则可以用列表来实现。

【例 3-18】 Id 为 101 的用户某时段的浏览记录为{a. html，b. html，…，g. html}，要求将用户最近的 5 条浏览记录保留 3 天。测试代码如文件 3-25 所示。

【文件 3-25】 例 3-18 测试代码

```
1    @Test
2    public void testViewed(){
3        String[] pages = {"a.html","b.html","c.html","d.html","e.html",
4            "f.html","g.html"};
5        //保留最近 5 条浏览记录
6        int viewed_page_counter = 5;
7        int offset = pages.length - viewed_page_counter;
8        for(int i = 0;i < pages.length;i++) {
9            if(i < offset) {
10               template.opsForList().rightPush("101:20241011:viewed",
11                   pages[i]);
12               template.opsForList().leftPop("101:20241011:viewed");
13           } else {
14               template.opsForList().leftPush("101:20241011:viewed",
15                   pages[i]);
16               template.expire("101:20241011:viewed",3 * 24 * 60,
17                   TimeUnit.MINUTES);
18           }
19       }
20       System.out.println("浏览记录为：");
21       List < String > viewedPages = template.opsForList().range(
22           "101:20241011:viewed",0, - 1);
23       viewedPages.forEach(System.out::println);
24   }
```

测试代码运行结果如图 3-12 所示。

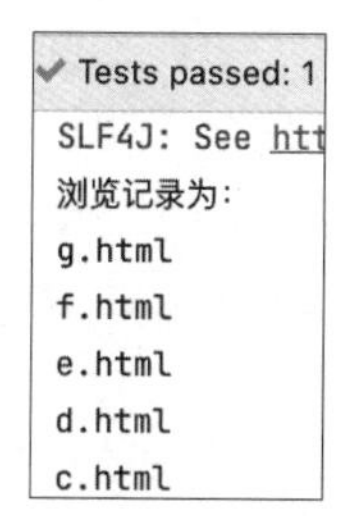

图 3-12 例 3-18 测试代码运行结果

3.5 Spring 操作 Redis 哈希

几乎所有的编程语言都提供了哈希类型。Redis 的哈希类型值是一个键值对结构，形如 value={{field_1,value_1},…,{field_n,value_n}}，因此哈希类型特别适合存储对象。Redis 是以键值对的形式存储数据的。Redis 键值对和哈希类型二者的关系可以用图 3-13 表示。

如图 3-13 所示，普通哈希类型数据< name，Tom >与< age，28 >以键 user:1 存储在 Redis 中。其映射关系在 Redis 中叫作字段值(field-value)，注意这里的值是指字段(field)对应的值，不是键对应的值。操作 Redis 哈希的方法与操作 Redis 字符串的方法类似，要调用 RedisTemplate 类的 opsForHash()方法创建 HashOperations 子接口对象，再调用 HashOperations 子接口的相关方法。本节介绍 HashOperations 子接口(org.springframework.data.redis.core.HashOperations < K，V >)中的主要方法的使用。

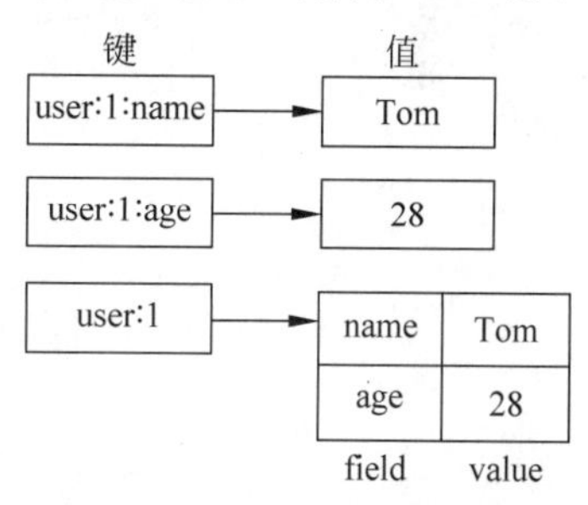

图 3-13 Redis 键值对和哈希类型的关系

(1) 方法原型：void put(H key，HK hashKey，HV value)；功能：将哈希 key 中的字段 hashKey 的值设置为 value；对应的 Redis 命令：HSET。

(2) 方法原型：@Nullable HV get(H key，Object hashKey)；功能：从哈希 key 中根据字段 hashKey 取出值；对应的 Redis 命令：HGET。

(3) 方法原型：Map < HK，HV > entries(H key)；功能：根据 key 获取整个哈希存储的值；对应的 Redis 命令：HGETALL。

(4) 方法原型：Set < HK > keys(H key)；功能：获取哈希 key 的所有字段名的集合；对应的 Redis 命令：HKEYS。

(5) 方法原型：List < HV > values(H key)；功能：获取哈希 key 的所有字段的值；对应的 Redis 命令：HVALS。

(6) 方法原型：Long size(H key)；功能：返回哈希中字段的数量；对应的 Redis 命令：HLEN。

(7) 方法原型：Boolean hasKey(H key，Object hashKey)；功能：判断哈希 key 中给定的字段 hashKey 是否存在；对应的 Redis 命令：HEXISTS。

【例 3-19】 Redis 哈希基础操作 1。要求：①将哈希数据< name:Tom >、< age:26 >、

<class:3>存储到哈希 rHash 中；②返回 rHash 中字段的数量；③取出年龄值；④取出 rHash 中存储的全部值；⑤取出全部字段名；⑥取出全部字段值；⑦判断 rHash 中是否存在 age 字段和 ttt 字段。测试代码如文件 3-26 所示。

【文件 3-26】 例 3-19 测试代码

```
1   @Test
2   public void testEx17() {
3       //①
4       template.opsForHash().put("rHash", "name", "Tom");
5       template.opsForHash().put("rHash", "age", "26");
6       template.opsForHash().put("rHash", "class", "6");
7       //②
8       assertEquals(3, template.opsForHash().size("rHash").intValue());
9       //③
10       assertEquals("26", template.opsForHash().get("rHash", "age"));
11      //④
12      System.out.println("stored Hash:" +
13          template.opsForHash().entries("rHash"));
14      //⑤
15      System.out.println("fields:" +
16          template.opsForHash().keys("rHash"));
17      //⑥
18      System.out.println("values:" +
19          template.opsForHash().values("rHash"));
20      //⑦
21      assertTrue(template.opsForHash().hasKey("rHash", "age"));
22      assertFalse(template.opsForHash().hasKey("rHash", "ttt"));
23  }
```

```
stored Hash:{name=Tom, age=26, class=6}
fields:[class, name, age]
values:[6, Tom, 26]
```

图 3-14 例 3-19 测试代码运行结果

运行上述测试代码,结果如图 3-14 所示。

(8) 方法原型：Cursor<Map.Entry<HK,HV>> scan(H key,ScanOptions options)；功能：用于增量迭代哈希 key 中的键值对,参数 ScanOptions 是用于 SCAN 命令的选项,目前的常量值为 NONE(对扫描模式不做限制)；对应的 Redis 命令：HSCAN。

该命令支持增量迭代,即每次执行都只会返回少量元素,所以该命令可以用于生产环境,而不会出现像 KEYS 命令带来的问题：当 KEYS 命令被用于处理一个大的数据库时,可能会阻塞服务器达数秒之久。在对键进行增量式迭代的过程中,键可能会被修改,所以增量式迭代命令只能对被返回的元素提供有限的保证。

(9) 方法原型：Long increment(H key,HK hashKey,long delta)；功能：为哈希 key 中的字段 hashKey 的值加上增量 delta。增量也可以为负数,相当于对给定字段进行减法操作。如果 key 不存在,则会创建一个新的哈希并执行该方法。如果域 hashKey 不存在,那么在执行该方法前,字段的值被初始化为 0。对一个存储字符串值的字段执行该方法将造成一个错误。对应的 Redis 命令：HINCRBY。

由于内部序列化器的设置不同,因此此方法要求配合 StringRedisTemplate 类使用。

(10) 方法原型：void putAll(H key,Map<?extends HK,?extends HV>m)；功能：将

多个字段值对同时设置到哈希 key 中。执行此方法会覆盖哈希中已存在的字段。如果 key 不存在,则创建一个空的哈希并执行该方法;对应的 Redis 命令:HMSET。

(11) 方法原型:List <HV> multiGet(H key,Collection <HK> hashKeys);功能:返回哈希 key 中,一个或多个给定字段 hashKeys 的值。如果给定的字段不存在于哈希,则返回一个 nil 值。对应的 Redis 命令:HMGET。

(12) 方法原型:Long delete(H key,Object...hashKeys);功能:删除哈希 key 中一个或多个指定的字段 hashKeys。对应的 Redis 命令:HDEL。

【例 3-20】 Redis 哈希基础操作 2。要求:①使用 scan()方法遍历例 3-19 中的哈希 rHash,并输出其全部值;②将 age 字段的值增加 1;③将哈希数据<name:Bob>、<age:28>、<class:2>一次性加入哈希 rHash2 中;④从哈希 rHash 中取出 name 字段和 age 字段的值;⑤删除哈希 rHash 中的字段 name。测试代码如文件 3-27 所示。

【文件 3-27】 例 3-20 测试代码

```
1   @Test
2   public void testEx18 () {
3       //①
4       template.opsForHash().put("rHash", "name", "Tom");
5       template.opsForHash().put("rHash", "age", "26");
6       template.opsForHash().put("rHash", "class", "6");
7       System.out.println("all the values in rHash are:");
8       Cursor < Map.Entry < Object,Object >> cursor =
9            template.opsForHash().scan("rHash", ScanOptions.NONE);
10      cursor.forEachRemaining(entry -> System.out.println(
11           entry.getKey() + ":" + entry.getValue()));
12      //②
13      //需注入 StringRedisTemplate
14      assertEquals("27",template.opsForHash().increment(
15          "rHash","age",1).toString());
16      //③
17      Map < String,Object > tempMap = new HashMap < String,Object >();
18      tempMap.put("name","Bob");
19      tempMap.put("age","28");
20      tempMap.put("class","2");
21      template.opsForHash().putAll("rHash2",tempMap);
22      System.out.println("rHash:"
23          + template.opsForHash().entries("rHash"));
24      System.out.println("rHash2:"
25          + template.opsForHash().entries("rHash2"));
26      //④
27      List < Object > ks = new ArrayList < Object >();
28      ks.add("name");
29      ks.add("age");
30      System.out.println("values for field name and age for rHash:");
31      System.out.println(template.opsForHash().multiGet("rHash",ks));
32      //⑤
33      template.opsForHash().delete("rHash","name");
34      System.out.println("name field has been removed. Now rHash looks
35         like this: " + template.opsForHash().entries("rHash"));
36  }
```

运行上述测试代码，结果如图 3-15 所示。

```
all the values in rHash are:
name:Tom
class:6
age:26
----------
rHash:{age=26, class=6, name=Tom}
rHash2:{age=28, name=Bob, class=2}
----------
values for field name and age for rHash:
[Tom, 26]
----------
name field has been removed. Now rHash looks like this: {class=6, age=26}
```

图 3-15　例 3-20 测试代码运行结果

目前的软件系统（包括电商和社交网络）经常有这样的需求：根据用户提供的手机号码发送验证码，实现登录。下面的例子在例 3-9 的基础上模拟手机验证码的登录验证功能。

【例 3-21】 模拟手机验证码登录功能，用户提交手机上收到的验证码并完成登录验证。对于验证成功的用户，将其登录信息保存到 Redis 中，测试代码如文件 3-28 所示。

【文件 3-28】 UserLogin. java

```
1    public class UserLogin {
2        public String login(StringRedisTemplate template, String phone,
3            String userCode){
4            String codeKey = phone + "_CODE";
5            String cacheCode = template.opsForValue().get(codeKey);
6            if(cacheCode == null || ! cacheCode.equals(userCode)) {
7                System.out.println("验证码错误");
8                return "FAIL";
9            }
10           User user = new User();
11           user.setUserId(Integer.valueOf(133));
12           user.setName("admin");
13           user.setPhone(phone);
14           String token = UUID.randomUUID().toString();
15           Map<String,Object> map = BeanUtil.beanToMap(user,
16               new HashMap<>(), CopyOptions.create()
17               .setIgnoreNullValue(true)
18               .setFieldValueEditor((fieldName,fieldValue)
19               -> fieldValue.toString()));
20           System.out.println(map);
21           String tokenKey = phone + "_" + token;
22           template.opsForHash().putAll(tokenKey,map);
23           template.expire(tokenKey,30, TimeUnit.MINUTES);
24           template.delete(codeKey);
25           return "OK";
26       }
27   }
```

如文件 3-28 所示，本例在例 3-9 基础上完成登录验证任务。第 4 行指定 Redis 中缓存的已发送给用户的验证码的键 codeKey。第 5 行根据 codeKey 获取发送给用户的验证码。第 6～9 行将已发送给用户的验证码和用户提交的验证码进行比对，若比对失败则报告错误并退出。第 10～13 行先实例化 User 类(代码见本书配套源代码),再对其属性分别赋值。这些操作用来模拟通过手机号在数据库中检索用户的相关信息。第 14 行生成随机令牌,用来保存用户会话(Session)信息。第 15～19 行将 User 类的对象 user 转换为 Map,目的是准备将 Map 对象存入 Redis 哈希中。其中的 BeanUtil 是 hutool 提供的工具类,引入的依赖见例 3-9。第 22 行将用户信息存入 Redis 哈希。第 23 行设置用户信息的过期时间。用户登录验证成功,删除缓存的用户验证码(第 24 行)。

本例中,在用户登录验证成功后,将用户信息写入 Redis。在 Web 应用系统开发中,用户登录验证成功后,通常将用户信息写入 Session。当 Web 应用系统部署在一台 Tomcat 服务器上时,这样做是可行的。而工程上,为应对大量的并发请求,往往将 Web 应用系统部署到 Tomcat 集群上。而一个 Session 对象不能在多个 Tomcat 上使用,这样用户登录信息只能保存在一个 Tomcat 上。因为集群的存在,用户的请求会被分配到不同的 Tomcat 上处理。这样,当登录过的用户再次向系统发送请求时,请求可能会被分配到没有保存用户信息的 Tomcat。这时,该 Tomcat 就会要求用户重新登录,这样就会极大降低用户的体验度。并且,Tomcat 集群中也会保存大量冗余的用户登录信息,造成资源浪费。本例中,将用户的登录信息保存到 Redis 中就是对上述问题的一个解决方案。一方面,Tomcat 集群可以从 Redis 中获取用户的登录信息,实现数据共享;另一方面,由于 Redis 具有良好的读写性能,可以从容应对众多用户登录时带来的大量的并发请求。同时,本例给出的解决方案还有一个不足,就是第 23 行中设置了用户信息的保存时间为 30 分钟。如果用户的操作时长超过 30 分钟,会因为 Redis 中缓存的信息过期而被要求重新登录。这个问题需要其他技术手段来解决,此处不再详述。

随后,测试用户的登录验证功能,在文件 3-16 的基础上增加一个测试用例,测试代码如文件 3-29 所示。

【文件 3-29】 增加的测试用例

```
1   @Test
2   public void testLogin(){
3       UserLogin userLogin = new UserLogin();
4       userLogin.login(template,"15612345678","******");
5   }
```

测试时,需要先接收验证码。因此,要先执行文件 3-16 中的测试用例,将验证码填写到文件 3-29 第 4 行的“*”处。如果验证成功,则可利用 Redis 客户端查看 Redis 中保存的用户信息。Redis 中保存的用户信息及程序运行结果如图 3-16 所示。

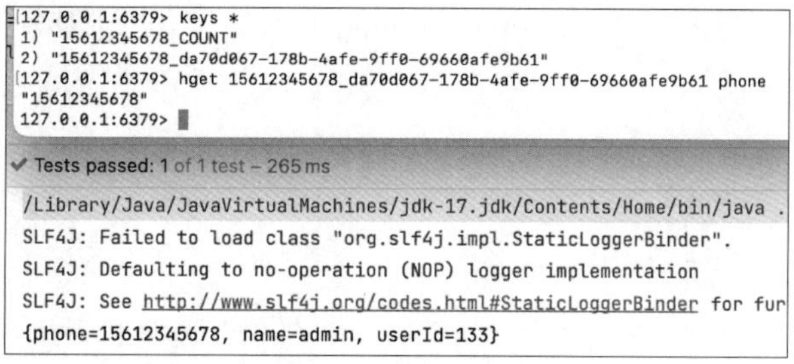

```
127.0.0.1:6379> keys *
1) "15612345678_COUNT"
2) "15612345678_da70d067-178b-4afe-9ff0-69660afe9b61"
127.0.0.1:6379> hget 15612345678_da70d067-178b-4afe-9ff0-69660afe9b61 phone
"15612345678"
127.0.0.1:6379>
✔ Tests passed: 1 of 1 test – 265 ms
/Library/Java/JavaVirtualMachines/jdk-17.jdk/Contents/Home/bin/java .
SLF4J: Failed to load class "org.slf4j.impl.StaticLoggerBinder".
SLF4J: Defaulting to no-operation (NOP) logger implementation
SLF4J: See http://www.slf4j.org/codes.html#StaticLoggerBinder for fur
{phone=15612345678, name=admin, userId=133}
```

图 3-16 Redis 中保存的用户信息及程序运行结果

3.6 Spring 操作 Redis 集合

与列表类型数据相似，集合类型数据也可以在同一个键下存储一个或多个元素。与列表类型不同的是，集合中不允许有重复元素，并且集合中的元素是无序的，不能通过索引获取元素。Redis 的集合是字符串类型元素的无序集合。Redis 除了支持集合内的增、删、改、查操作，还支持多个集合的交、并、差运算。

操作 Redis 集合的方法与操作 Redis 字符串的方法类似，要调用 RedisTemplate 类的 opsForSet()方法创建 SetOperations 子接口对象，再调用 SetOperations 子接口的相关方法。本节介绍 SetOperations 子接口(org.springframework.data.redis.core.SetOperations < K,V >)中的主要方法的使用。

(1) 方法原型：@Nullable Long add(K key,V...values)；功能：向集合 key 中添加元素，并返回添加的个数；对应的 Redis 命令：SADD。

(2) 方法原型：@Nullable Boolean move(K key,V value,K destKey)；功能：将元素 value 从集合 key 移动到集合 destKey；对应的 Redis 命令：SMOVE。

(3) 方法原型：@Nullable Set < V > members(K key)；功能：返回集合 key 中的所有成员，不存在的 key 被视为空集合；对应的 Redis 命令：SMEMBERS。

(4) 方法原型：@Nullable List < V > pop(K key,long count)；功能：从集合 key 中删除并返回 count 个随机成员；对应的 Redis 命令：SPOP。

(5) 方法原型：@Nullable Long remove(K key,Object...values)；功能：移除集合 key 中的一个或多个元素 value，不存在的元素 value 会被忽略。当 key 不是集合类型时，返回一个错误；对应的 Redis 命令：SREM。

(6) 方法原型：@Nullable Long size(K key)；功能：返回集合 key 的基数(集合中元素的数量)；对应的 Redis 命令：SCARD。

(7) 方法原型：@Nullable Boolean isMember(K key,Object o)；功能：检测集合 key 是否包含元素 o；对应的 Redis 命令：SISMEMBER。

(8) 方法原型：Cursor < V > scan(K key,ScanOptions options)；功能：支持增量式迭代集合中的元素；对应的 Redis 命令：SSCAN。

【例 3-22】 给定两个集合 ball 和 ball2，对两个集合执行下述操作：①向集合 ball 中添加 4 个元素，分别为 football、volleyball、basketball 和 pingpong；②输出集合 ball 中的所有元素；③迭代输出集合 ball 中的元素；④将集合 ball 中的元素 pingpong 移至集合 ball2 中，检查集合 ball 中元素的个数，并检查元素 basketball 是否属于集合 ball；⑤从集合 ball 中随机取出两个元素并输出；⑥将元素 pingpong 从集合 ball2 中移除并输出集合 ball2 的剩余元素。测试代码如文件 3-30 所示。

【文件 3-30】 例 3-22 测试代码

```
1    @Test
2    public void testEx19() {
3        //①
4        String[] s = {"football","volleyball","basketball","pingpong"};
5        template.opsForSet().add("ball",s);
```

```
6        //②
7        System.out.println("the members returned by members() method:");
8        System.out.println(template.opsForSet().members("ball"));
9        //③
10       System.out.println("----------------");
11       System.out.println("the members returned by scan() method:");
12       template.opsForSet().scan("ball", ScanOptions.NONE)
13          .forEachRemaining(System.out::println);
14       //④
15       template.opsForSet().move("ball","pingpong","ball2");
16       assertEquals(3,template.opsForSet().size("ball").intValue());
17       assertTrue(template.opsForSet().isMember("ball","basketball"));
18       //⑤
19       System.out.println("----------------");
20       System.out.println("obtain 2 items randomly from ball set");
21       System.out.println(template.opsForSet().pop("ball",2));
22       //⑥
23       System.out.println("----------------");
24       System.out.println("remove pingpong from ball2 set");
25       template.opsForSet().remove("ball2","pingpong");
26       System.out.println("the remaining elements in ball2 set :"
27           + template.opsForSet().members("ball2"));
28   }
```

运行上述测试代码,运行结果如图 3-17 所示。

```
the members returned by members() method:
[basketball, football, volleyball, pingpong]
----------------
the members returned by scan() method:
basketball
football
volleyball
pingpong
----------------
obtain 2 items randomly from ball set
[football, basketball]
----------------
remove pingpong from ball2 set
the remaining elements in ball2 set :[]
```

图 3-17 例 3-22 测试代码运行结果

(9) 方法原型:@Nullable Set < V > intersect(K key,Collection < K > otherKeys);功能:求集合 key 与其他多个集合 otherKeys 的交集,不存在的 key 被视为空集。该方法的另外两种重载形式:@Nullable Set < V > intersect(K key,K otherKey)和@Nullable Set < V > intersect(Collection < K > keys);对应的 Redis 命令:SINTER。

(10) 方法原型:@Nullable Long intersectAndStore(K key,K otherKey,K destKey);功能:求集合 key 与集合 otherKey 的交集,并将产生的交集元素存入集合 destKey 中。如果集合 destKey 已经存在,则将其覆盖。集合 destKey 可以是集合 key 本身。该方法的另外两种重载形式:@Nullable Long intersectAndStore(K key,Collection < K > otherKeys,K destKey)和@Nullable Long intersectAndStore(Collection < K > keys,K destKey);对应

的 Redis 命令：SINTERSTORE。

（11）方法原型：@Nullable Set < V > union(K key,K otherKey)；功能：求集合 key 与集合 otherKey 的并集，不存在的 key 被视为空集。该方法的另外两种重载形式：@Nullable Set < V > union(K key,Collection < K > otherKeys)和@Nullable Set < V > union(Collection < K > keys)；对应的 Redis 命令：SUNION。

（12）方法原型：@Nullable Long unionAndStore(K key,K otherKey,K destKey)；功能：求集合 key 和集合 otherKey 的并集，并将产生的并集元素存入集合 destKey 中。如果集合 destKey 已经存在，则将其覆盖。集合 destKey 可以是集合 key 本身。该方法的另外两种重载形式：@Nullable Long unionAndStore(K key,Collection < K > otherKeys,K destKey)和@Nullable Long unionAndStore(Collection < K > keys,K destKey)；对应的 Redis 命令：SUNIONSTORE。

（13）方法原型：@Nullable Set < V > difference(K key,K otherKey)；功能：求集合 key 与集合 otherKey 的差集，不存在的 key 被视为空集。该方法的另外两种重载形式：@Nullable Set < V > difference(K key,Collection < K > otherKeys)和@Nullable Set < V > difference(Collection < K > keys)；对应的 Redis 命令：SDIFF。

（14）方法原型：@Nullable Long differenceAndStore(K key,K otherKey,K destKey)；功能：求集合 key 和集合 otherKey 的差集，并将产生的差集元素存入集合 destKey 中。如果集合 destKey 已经存在，则将其覆盖。集合 destKey 可以是集合 key 本身。该方法的另外两种重载形式：@Nullable Long differenceAndStore(K key,Collection < K > otherKeys,K destKey)和@Nullable Long differenceAndStore(Collection < K > keys,K destKey)；对应的 Redis 命令：SDIFFSTORE。

【例 3-23】 在利用 Redis 存储社交关系时，可能会有如下需求：①在微博中 zhangsan 有一批好友，lisi 有另外一批好友，现需要查询 zhangsan 和 lisi 的共同好友；②要得到 zhangsan、lisi 和 wangwu 关注的所有公众号；③要得到 zhangsan 关注的但 lisi 和 wangwu 没有关注的公众号。测试代码如文件 3-31 所示。

【文件 3-31】 例 3-23 测试代码

```
1   @Test
2   public void testEx20(){
3       //准备好友数据
4       template.opsForSet().add("zhangsan-friend",
5           "friend1","friend2","friend3");
6       template.opsForSet().add("lisi-friend",
7           "friend1","friend3","friend4");
8       //①
9       System.out.println("Mutual friends of zhangsan & lisi:"
10          + template.opsForSet().intersect(
11      "zhangsan-friend","lisi-friend"));
12      //准备公众号数据
13      template.opsForSet().add("zhangsan-concern",
14          "pub1","pub2","pub3","pub4");
15      template.opsForSet().add("lisi-concern",
16          "pub1","pub2","pub5","pub6");
17      template.opsForSet().add("wangwu-concern",
```

```
18              "pub1","pub7","pub8","pub9");
19          //②
20          System.out.println("All concerns are :");
21          List<String> list = new ArrayList<String>();
22          list.add("lisi-concern");
23          list.add("wangwu-concern");
24          template.opsForSet().union("zhangsan-concern",list).iterator()
25            .forEachRemaining(e->System.out.print(e+" "));
26          System.out.println();
27          //③
28          template.opsForSet().differenceAndStore("zhangsan-concern",
29             list,"zhangsanonly");
30          System.out.println("only zhangsan's concern :"
31             +template.opsForSet().members("zhangsanonly"));
32      }
```

如文件3-31所示，第4～7行准备zhangsan和lisi的好友数据。两者的共同好友可以通过求解两者好友集合的交集得到(第9～11行)。第13～18行准备zhangsan、lisi和wangwu的关注公众号数据。要得到三个人关注的所有公众号，只要求解三者关注公众号集合的并集即可。在求解并集时，调用的是union()方法的一种重载形式：@Nullable Set<V>union(K key,Collection<K>otherKeys)(第24行)。最后遍历结果集合，输出并集的全部元素(第25、26行)。要得到只有zhangsan关注的公众号，只需求解zhangsan关注的公众号集合与其他两人关注的公众号集合的差集即可(第28、29行)。运行此测试代码，运行结果如图3-18所示。

```
Mutual friends of zhangsan & lisi:[friend3, friend1]
All concerns are :
pub1 pub4 pub6 pub5 pub2 pub3 pub7 pub8 pub9
only zhangsan's concern :[pub3, pub4]
```

图3-18 例3-23测试代码运行结果

3.7 Spring操作Redis有序集合

有序集合(Sorted set，也称ZSet)同集合有一定的相似性，也是字符串类型元素的集合，并且都不能出现重复元素。在有序集合里，每个数据都会对应一个double类型的参数score(分数)。Redis正是按照分数来为集合中的元素进行升序排列。有序集合的元素是唯一的，但分数却可以重复。表3-4给出了Redis列表、集合和有序集合的异同点。

表3-4 Redis列表、集合和有序集合的异同点

数据结构	是否允许重复元素	是否有序	有序实现方式	应用场景
列表	是	是	索引(下标)	时间轴、消息队列等
集合	否	否	无	标签、社交网络等
有序集合	否	是	分数	排行榜、社交网络等

操作Redis有序集合的方法与操作Redis字符串的方法类似，要调用RedisTemplate类的opsForZSet()方法创建ZSetOperations子接口对象，再调用ZSetOperations子接口的相关方法。本节介绍ZSetOperations子接口(org.springframework.data.redis.core.ZSetOperations

<K,V>)中的主要方法的使用。

3.7.1 对单个集合的操作

(1) 方法原型：@Nullable Boolean add(K key,V value,double score)；功能：将一个或多个元素value及其分数score加入有序集合key中。如果某个元素value已经是有序集合的成员，那么更新这个元素value的分数score，并通过重新插入这个元素来保证其在正确的位置上。分数score可以是整数值或双精度浮点数。如果有序集合key不存在，则创建一个空的有序集并执行add()方法。当key存在但不是有序集合类型时，返回一个错误。该方法的另外一种重载形式：@Nullable Long add(K key,Set <ZSetOperations.TypedTuple <V>> tuples)；返回值：被成功添加的新成员的数量，不包括那些被更新的、已经存在的成员；对应的Redis命令：ZADD。

(2) 方法原型：@Nullable Set <V> range(K key,long start,long end)；功能：返回有序集合在指定区间[start,end]内的成员；下标参数start和stop都以0表示有序集合的第一个成员，以1表示有序集合的第二个成员，以此类推。也可以使用负数下标，以−1表示最后一个成员，−2表示倒数第二个成员，以此类推。超出范围的下标并不会引起错误。例如，当start的值比有序集合的最大下标还要大或start>end时，该方法只是简单地返回一个空列表。另外，假如参数end的值比有序集的最大下标还要大，那么Redis将end当作最大下标来处理。对应的Redis命令：ZRANGE。

(3) 方法原型：@Nullable Set <ZSetOperations.TypedTuple <V>> rangeWithScores(K key,long start,long end)；功能：返回有序集合key的下标在指定区间[start,end]内的成员对象，其中有序集成员按分数值递增顺序排列；对应的Redis命令：ZRANGE。

(4) 方法原型：@Nullable default Set <V> reverseRangeByLex(K key,RedisZSetCommands.Range range)；功能：从有序集合key中获取range范围内的具有反向字典顺序的所有元素；其中的RedisZSetCommands.Range对应的Redis命令：ZREVRANGEBYLEX。

(5) 方法原型：@Nullable Set <ZSetOperations.TypedTuple <V>> popMax(K key,long count)；功能：从有序集合key中返回并移除count个分数最大的元素。

(6) 方法原型：@Nullable Long zCard(K key)；功能：获取有序集合key的成员数；对应的Redis命令：ZCARD。

(7) 方法原型：@Nullable Long removeRange(K key,long start,long end)；功能：移除有序集合key中指定排名区间[start,end]内的所有成员。下标参数start和end都以0表示有序集合的第一个成员，1表示有序集合的第二个成员，以此类推。也可以使用负数下标，−1表示最后一个成员，−2表示倒数第二个成员，以此类推。返回值：被移除的成员数量；对应的Redis命令：ZREMRANGEBYRANK。

【例3-24】 有序集合基本操作1(模拟学生成绩管理)。要求：①向有序集合students中添加4个元素，元素名字分别为SuXun、SuShi、SuZhe、HanYu，分数分别为80.0、81.0、81.0、88.0；②输出students集合中的全部元素；③将students中的元素按分数由低到高排序并输出；④将students中的元素按分数由高到低排列，在分数相同的情况下，按反字母表序排列；⑤取出分数最高的元素；⑥删除students集合中的剩余元素。测试代码如文件3-32所示。

【文件 3-32】 例 3-24 测试代码

```
1   @Test
2   public void testEx21() {
3       ZSetOperations.TypedTuple<String> tuple1 =
4           new DefaultTypedTuple<String>("SuXun",80.0);
5       ZSetOperations.TypedTuple<String> tuple2 =
6           new DefaultTypedTuple<String>("SuShi",81.0);
7       ZSetOperations.TypedTuple<String> tuple3 =
8           new DefaultTypedTuple<String>("SuZhe",81.0);
9       ZSetOperations.TypedTuple<String> tuple4 =
10          new DefaultTypedTuple<String>("HanYu",88.0);
11      Set<ZSetOperations.TypedTuple<String>> t =
12          new HashSet<ZSetOperations.TypedTuple<String>>();
13      t.add(tuple1);    t.add(tuple2);
14      t.add(tuple3);    t.add(tuple4);
15      //①
16      template.opsForZSet().add("students",t);
17      //②
18      System.out.println(template.opsForZSet()
19          .range("students", -4, -1));
20      //③
21      System.out.println(template.opsForZSet()
22          .rangeWithScores("students",0, -1));
23      //④
24      RedisZSetCommands.Range range = new RedisZSetCommands.Range();
25      System.out.println(template.opsForZSet()
26          .reverseRangeByLex("students", range.gte("H")));
27      //⑤
28      System.out.println(template.opsForZSet().popMax("students",1));
29
30      assertEquals(3,template.opsForZSet()
31          .zCard("students").intValue());
32      //⑥
33      template.opsForZSet().removeRange("students",0, -1);
34      System.out.println(template.opsForZSet()
35          .range("students",0, -1));
36  }
```

其中,文件 3-32 的第 26 行调用 reverseRangeByLex()方法。该方法用于获取满足非分数的排序取值。这个排序只有在分数相同的情况下才能使用,如果有不同的分数则返回值不确定。运行此测试代码,运行结果如图 3-19 所示。

```
集合中的全部元素: [SuXun, SuShi, SuZhe, HanYu]
----------
分数由高到低排序输出:
[DefaultTypedTuple [score=80.0, value=SuXun],
 DefaultTypedTuple [score=81.0, value=SuShi],
 DefaultTypedTuple [score=81.0, value=SuZhe],
 DefaultTypedTuple [score=88.0, value=HanYu]]
----------
分数由高到低排序, 分数相同的情况下, 按字母表反序:
[HanYu, SuZhe, SuShi, SuXun]
----------
最高分:
[DefaultTypedTuple [score=88.0, value=HanYu]]
----------
移除集合中的所有元素:
[]
```

图 3-19 例 3-24 测试代码运行结果

(8) 方法原型:@Nullable Long count(K key, double min, double max);功能:返回分数在[min, max]区间内的成员个数;对应的 Redis 命令:ZCOUNT。

(9) 方法原型:@Nullable Long remove(K key, Object...values);功能:移除有序集合 key 中的一个

或多个成员，不存在的成员将被忽略。当 key 存在但不是有序集合类型时，返回一个错误。返回被成功移除的成员的数量，不包括被忽略的成员。对应的 Redis 命令：ZREM。

（10）方法原型：@Nullable Long rank(K key,Object o)；功能：返回有序集合 key 中指定成员 o 的排名，其中有序集合成员按分数值递增顺序排列；对应的 Redis 命令：ZRANK。

（11）方法原型：@Nullable Long reverseRank(K key,Object o)；功能：返回有序集合 key 中指定成员 o 的排名，其中有序集合成员按分数值递减顺序排列；对应的 Redis 命令：ZREVRANK。

（12）方法原型：@Nullable Set<ZSetOperations.TypedTuple<V>> reverseRangeWithScores(K key,long start,long end)；功能：返回有序集合 key 在指定区间[start,end]内的成员对象，其中有序集合成员按分数值递减顺序排列；对应的 Redis 命令：ZREVRANGEBYSCORE。

（13）方法原型：@Nullable Double score(K key,Object o)；功能：获取指定成员的 score 值；对应的 Redis 命令：ZSCORE。

（14）方法原型：@Nullable Set<V> distinctRandomMembers(K key,long count)；功能：从有序集合 key 中随机获取 count 个不同元素，该方法需要 Redis 6.2.0 及以上版本支持；对应的 Redis 命令：ZRANDMEMBER。

（15）方法原型：@Nullable Double incrementScore(K key,V value,double delta)；功能：将有序集合 key 中值为 value 的元素的分数增加增量 delta；对应的 Redis 命令：ZINCRBY。

（16）方法原型：Cursor<ZSetOperations.TypedTuple<V>> scan(K key,ScanOptions options)；功能：使用光标(Cursor)在有序集合 key 上迭代元素(包括元素成员和元素分数)；对应的 Redis 命令：ZSCAN。

【例 3-25】 有序集合基本操作 2(模拟视频网站的一些基础操作)。要求：①向有序集合 zset 中添加三个元素，元素名字分别为 zset-1、zset-2 和 zset-3，分数分别为 9.9、9.6、9.1；②对 zset 中的元素按分数升序排列并输出排序结果；③计算分数在 9.3 分以上的元素数量；④从集合 zset 中随机抽取一个元素，并获取其排名及分数；⑤将 zset-1 的分数提高到 10.1。测试代码如文件 3-33 所示。

【文件 3-33】 例 3-25 测试代码

```
1   @Test
2   public void testEx22() {
3       ZSetOperations.TypedTuple<String> objectTypedTuple1 =
4           new DefaultTypedTuple<String>("zset-1", 9.9);
5       ZSetOperations.TypedTuple<String> objectTypedTuple2 =
6           new DefaultTypedTuple<String>("zset-2", 9.6);
7       ZSetOperations.TypedTuple<String> objectTypedTuple3 =
8           new DefaultTypedTuple<String>("zset-3", 9.1);
9       Set<ZSetOperations.TypedTuple<String>> tuples =
10          new HashSet<ZSetOperations.TypedTuple<String>>();
11      tuples.add(objectTypedTuple1);
12      tuples.add(objectTypedTuple2);
13      tuples.add(objectTypedTuple3);
14      //①
```

```
15          template.opsForZSet().add("zset", tuples);
16          //②
17          tuples = template.opsForZSet().reverseRangeWithScores(
18              "zset", 0, -1);
19          Cursor cursor = template.opsForZSet()
20              .scan("zset", ScanOptions.NONE);
21          cursor.forEachRemaining(System.out::println);
22          //③
23          assertEquals(2, template.opsForZSet()
24              .count("zset", 9.3, 10.0).intValue());
25          //④
26          Set set = template.opsForZSet().distinctRandomMembers("zset", 1);
27          Iterator<String> iterator = set.iterator();
28          iterator.forEachRemaining(item -> System.out.println(item
29              + " 排名: " + String.valueOf(template.opsForZSet()
30              .rank("zset", item).intValue() + 1)
31              + " 分数: " + template.opsForZSet().score("zset", item)));
32          //⑤
33          assertEquals(10.1, template.opsForZSet()
34              .incrementScore("zset", "zset-1", 0.2).doubleValue(), 0.01);
35      }
```

运行测试代码，运行结果如图 3-20 所示。

```
按分数升序排列的结果:
DefaultTypedTuple [score=9.1, value=zset-3]
DefaultTypedTuple [score=9.6, value=zset-2]
DefaultTypedTuple [score=9.9, value=zset-1]
----------
随机抽取一个元素，排名及分数:
zset-2 排名: 11 分数: 9.6
----------
```

图 3-20　例 3-25 测试代码运行结果

3.7.2　对多个集合的操作

（1）方法原型：@Nullable Long unionAndStore(K key, Collection < K > otherKeys, K destKey, RedisZSetCommands. Aggregate aggregate, RedisZSetCommands. Weights weights)；功能：计算有序集合 key 和 otherKeys 的并集，并将结果存储于集合 destKey 中。其中，参数 aggregate 表示并集选项，默认值为 SUM，表示结果集合 destKey 中元素的分数为该元素在各集合中分数的和；参数 aggregate 的另两个值为 MAX 和 MIN，分别表示结果集合 destKey 中的元素的分数为该元素在各集合中分数的最大和最小值。参数 weights 用于指定若干乘数因子，参与并集运算的每个集合中的元素的分数与指定的乘法因子相乘，得到该元素在结果集合中的分数，该参数的默认值为 1。对应的 Redis 命令：ZUNIONSTORE。

（2）方法原型：@Nullable Set < ZSetOperations. TypedTuple < V >> unionWithScores(K key, Collection < K > otherKeys, RedisZSetCommands. Aggregate aggregate, RedisZSetCommands weights)；功能：计算两个有序集合的并集，其中参数 aggregate 和 weights 的含义同(1)，该方法需要 Redis 6.2.0 及以上版本支持；对应的 Redis 命令：ZUNION。

（3）方法原型：@Nullable Long intersectAndStore(K key, Collection < K > otherKeys,

K destKey,RedisZSetCommands. Aggregate aggregate,RedisZSetCommands. Weights weights)；功能：计算有序集合 key 和 otherKeys 的交集，并将结果存储于集合 destKey 中，其中参数 aggregate 和 weights 的含义同(1)；对应的 Redis 命令：ZINTERSTORE。

(4) 方法原型：@Nullable Set < ZSetOperations. TypedTuple < V >> intersectWithScores(K key,Collection < K > otherKeys,RedisZSetCommands. Aggregate aggregate,RedisZSetCommands Weights weights)；功能：计算两个有序集合的交集，其中参数 aggregate 和 weights 的含义同(1)；该方法需要 Redis 6.2.0 及以上版本支持；对应的 Redis 命令：ZINTER。

(5) 方法原型：@Nullable Long differenceAndStore(K key,Collection < K > otherKeys, K destKey)；功能：求解有序集合 key 和 otherKeys 的差集，并将结果存放于集合 destKey 中，该方法需要 Redis 6.2.0 及以上版本支持；对应的 Redis 命令：ZDIFFSTORE。

(6) 方法原型：@Nullable Set < ZSetOperations. TypedTuple < V >> differenceWithScores (K key,Collection < K > otherKeys)；功能：计算两个有序集合的差集，该方法需要 Redis 6.2.0 及以上版本支持；对应的 Redis 命令：ZDIFF。

【例 3-26】 公司决定调整岗位工资。目前的岗位情况如下：LiBai、XinQiji 为创作岗位，岗位工资分别为 2300、2100；YueFei、XinQiji 为管理岗位，岗位工资分别为 3300 和 3700。要求：①查看只供职于创作岗位的人员及其岗位工资；②查看供职于多个岗位的人员及其最高档位的岗位工资；③对所有人员的岗位工资普遍调整，创作岗位人员在原岗位工资基础上翻倍，管理岗位人员在原岗位工资基础上上浮 20%，对兼任多个岗位的人员，岗位工资按各自岗位工资调整细则调整后就高，输出岗位工资调整后的所有人员的岗位工资。测试代码如文件 3-34 所示。

【文件 3-34】 例 3-26 测试代码

```
1    @Test
2    public void testEx23(){
3        ZSetOperations.TypedTuple < String > p1 =
4            new DefaultTypedTuple < String >("LiBai", 2300.0);
5        ZSetOperations.TypedTuple < String > p2 =
6            new DefaultTypedTuple < String >("XinQiji", 2100.0);
7        Set < ZSetOperations.TypedTuple < String >> ptuples =
8            new HashSet < ZSetOperations.TypedTuple < String >>();
9        ptuples.add(p1);        ptuples.add(p2);
10       template.opsForZSet().add("poet",ptuples);
11       ZSetOperations.TypedTuple < String > m1 =
12           new DefaultTypedTuple < String >("YueFei", 3300.0);
13       ZSetOperations.TypedTuple < String > m2 =
14           new DefaultTypedTuple < String >("XinQiji", 3700.0);
15       Set < ZSetOperations.TypedTuple < String >> mtuples =
16           new HashSet < ZSetOperations.TypedTuple < String >>();
17       mtuples.add(m1);        mtuples.add(m2);
18       template.opsForZSet().add("general",mtuples);
19       List < String > sets = new ArrayList < String >();
20       sets.add("general");
21       //①
22       System.out.println("创作岗位人员：");
23       template.opsForZSet().differenceAndStore("poet",sets,"res");
24       System.out.println(template.opsForZSet()
```

```
25              .rangeWithScores("res",0, - 1));
26          //②
27          System.out.println("兼任多个岗位人员: ");
28          System.out.println(template.opsForZSet().intersectWithScores(
29              "poet",sets,RedisZSetCommands.Aggregate.MAX));
30          //③
31          System.out.println("普调后的岗位工资: ");
32          template.opsForZSet().unionAndStore("poet",sets,"person",
33              RedisZSetCommands.Aggregate.MAX,
34              RedisZSetCommands.Weights.of(2.0,1.2));
35          System.out.println(template.opsForZSet()
36              .rangeWithScores("person",0, - 1));
37      }
```

其中,文件 3-34 的第 3～20 行为准备数据阶段。对于要求①,通过计算集合间的差集可求解(第 23 行)。对于要求②,只要求解两类人员集合的交集即可。第 28、29 行在调用 intersectWithScores()方法求解时,指定了参数 aggregate 的值为 MAX。对于要求③,基本思路是求解并集,但在执行并集操作前,需要将两个集合中的元素的分数分别与两个乘法因子相乘,第 34 行调用了 RedisZSetCommands.Weights 类的静态方法 of()为每个参与并运算的集合分别设置乘法因子。其中,of()方法的原型如下:

```
public static RedisZSetCommands.Weights of(double... weights)
```

运行测试代码,运行结果如图 3-21 所示。

```
创作岗位人员:
[DefaultTypedTuple [score=2300.0, value=LiBai]]
兼任多个岗位人员:
[DefaultTypedTuple [score=3700.0, value=XinQiji]]
普调后的岗位工资:
[DefaultTypedTuple [score=3960.0, value=YueFei],
 DefaultTypedTuple [score=4440.0, value=XinQiji],
 DefaultTypedTuple [score=4600.0, value=LiBai]]
```

图 3-21 例 3-26 测试代码运行结果

3.8 Spring 操作 HyperLogLog

HyperLogLog(超级日志,以下简称 HLL)是 Redis 2.8.9 版本增加的数据结构,它可以在不保存集合元素的情况下进行集合基数(集合中元素的个数)的统计。在 Redis 中,每个 HLL 键只需要花费 12KB 内存,就可以计算接近 2^{64} 个不同元素的基数。HLL 的工作原理是 HLL 概率算法。HLL 的统计规则是基于概率的,不会非常准确,其标准差为 0.81%。这个误差在工程上是完全可以接受的。

对于 HLL 的应用,通常有这样的场景:如要统计今天连接到网站的 IP 地址的数量,同一个 IP 多次访问计数为 1。此问题可抽象化:如原始的数据记录为{1,1,2,3,3,3,3,4,4,5},去重后的数据访问记录集合为 Visitors={1,2,3,4,5},Visitors 集合的基数为 5。对于计算集合基数的问题,一般可采用集合类型(Java 和 Redis 中都有集合类型)。对于数据量小的应用场景来说,这是可行的。但是随着数据量增大,集合只会无限扩张,最后变得很庞大,占用大量内存空间。此时，可以用 HLL 来解决问题。因为 HLL 只会根据输入元素来计算基数，而不会存储元素本身。

操作 HLL 的方法与操作 Redis 字符串的方法类似，要调用 RedisTemplate 类的 opsForHyperLogLog( ) 方法创建 HyperLogLogOperations 子接口对象，再调用 HyperLogLogOperations 子接口的相关方法。本节介绍 HyperLogLogOperations < K,V > 子接口(org. springframework. data. redis. core. HyperLogLogOperations < K,V >)中的主要方法的使用。

(1) 方法原型：Long add(K key,V...values)；功能：将给定的一个或多个值 values 添加到键 key。返回值：至少有一个值添加到键 key 时返回 1，否则返回 0；在流水线或事务中使用时返回 null。对应的 Redis 命令：PFADD。

(2) 方法原型：void delete(K key)；功能：删除给定的键 key。

(3) 方法原型：Long size(K...keys)；功能：获取键 keys 中当前元素的数量，在流水线或事务中使用时返回 null。对应的 Redis 命令：PFCOUNT。

(4) 方法原型：Long union(K destination,K...sourceKeys)；功能：将键 sourceKeys 的所有值合并到键 destination 中；对应的 Redis 命令：PFMERGE。

【例 3-27】 分别以 u1 和 u2 为键，向 HyperLogLog 各存入 10 000 000(1000 万)个范围在[0,99 999]的随机整数。输出 u1 和 u2 中元素的个数，并将 u1 和 u2 合并为 u，输出 u 中元素的个数。最后删除 u1、u2 和 u。测试代码如文件 3-35 所示。

【文件 3-35】 例 3-27 测试代码

```
1     @Test
2     public void testHyperLogLogOperations() {
3        HyperLogLogOperations < String, String > hyperLogLog =
4      template.opsForHyperLogLog();
5        String[] r = new String[5000];
6        String[] s = new String[5000];
7        for(int i = 0;i < 10000000;i++) {
8           int p = i % 5000;
9           r[p] = String.valueOf((int)(Math.random() * 100000));
10          if(p == 4999)
11             hyperLogLog.add("u1",r);
12       }
13       for(int j = 0;j < 10000000;j++){
14          int q = j % 5000;
15          s[q] = String.valueOf((int)(Math.random() * 100000));
16          if(q == 4999)
17             hyperLogLog.add("u2",s);
18       }
19       System.out.println("the size of u1 is " + hyperLogLog.size("u1"));
20       System.out.println("the size of u2 is " + hyperLogLog.size("u2"));
21
22       hyperLogLog.union("u","u1","u2");
23       System.out.println("the size of union is " + hyperLogLog.size("u"));
24       /*
25       hyperLogLog.delete("u1");
26       hyperLogLog.delete("u2");
```

```
27        hyperLogLog.delete("u");
28        */
29    }
```

在运行此测试代码前，可利用 info memory 命令查看 Redis 的内存占用情况，如图 3-22 所示。

```
127.0.0.1:6379> info memory
# Memory
used_memory:725368
used_memory_human:708.37K
```

图 3-22 运行测试代码前 Redis 的内存占用情况

运行测试代码后，控制台输出如图 3-23 所示。从输出结果可知，HLL 能够实现元素的自动去重，效果和集合类似。此外，union()方法（第 22 行）将两个 HLL 进行了合并，并且去掉了重复的元素，然后返回不重复元素个数。再次使用 info memory 命令检查 Redis 的内存占用情况，如图 3-24 所示。从代码运行前后的内存占用情况来看，向 Redis 中写入两千万条数据后（97 781 490B，约 93.25MB），内存占用由 725 368B 增加至 771 352B，只增加了 44.9KB。

```
the size of u1 is 99565
the size of u2 is 99565
the size of union is 99565
```

图 3-23 例 3-27 测试代码运行结果

```
127.0.0.1:6379> info memory
# Memory
used_memory:771352
used_memory_human:753.27K
```

图 3-24 运行测试代码后 Redis 的内存占用情况

HLL 的一项重要应用就是计数器。对于 Web 应用程序来讲，持续收集信息是一件非常重要的事。例如，已知网站在 5 分钟内得到 11 000 次点击，数据库在 5 秒内处理了 200 次写入和 600 次读取请求。这些数据都是非常重要的。因为通过在一段时间内持续地记录这些信息，可以了解网站流量的增减情况，进而根据这些数据预测何时需要对服务器进行升级，从而防止系统因为负载增加而宕机。对于网站来说，与流量相关的指标点有 UV（Unique Visitor，限定时段内同一客户端多次访问计为 1 次）、IP（用户的 IP 地址，限定时段内同一 IP 地址多次访问计为 1 次）等。

【例 3-28】 要求记录 0～24 时访问网站的 IP 地址的数量（同一 IP 地址多次访问计为 1 次），记录保留 24 小时。如果选用集合作为存储 IP 地址的数据类型，则会因为 IP 地址数量的增加导致集合占用的内存空间不断增大。本例只需要统计去重后的 IP 地址的数量，而无须记录每个 IP 地址，因此可以采用 HLL 来完成此任务。测试代码如文件 3-36 所示。

【文件 3-36】 例 3-28 测试代码

```
1    @Test
2    public void testIPCounter(){
3        HyperLogLogOperations<String, String> hLLOps =
4            template.opsForHyperLogLog();
5        String remoteAddr = "",backAddr = "",date_key = "";
6        Random random = new Random();
7        StringBuilder ipBuilder = new StringBuilder();
8        for(int j = 0; j < 10; j++) {
9            for (int i = 0; i < 4; i++) {
10               ipBuilder.append(random.nextInt(256));
11               if (i != 3)
12                   ipBuilder.append(".");
13           }
14           if( j < 8 ) {
```

```
15              remoteAddr = ipBuilder.toString();
16              if(j == 1)
17                  backAddr = remoteAddr;
18          }
19          else
20              remoteAddr = backAddr;
21          System.out.println("Remote IP : " + remoteAddr);
22          date_key = new SimpleDateFormat(
23              "yyyy - MM - dd HH:mm").format(System.currentTimeMillis());
24          hLLOps.add(date_key,remoteAddr);
25          template.expire(date_key,24,TimeUnit.HOURS);
26          ipBuilder.setLength(0);
27      }
28      System.out.println("IP 地址数: " + hLLOps.size(date_key));
29  }
```

```
Remote IP : 201.104.221.132
Remote IP : 110.124.243.96
Remote IP : 164.253.37.25
Remote IP : 61.150.245.217
Remote IP : 223.125.92.78
Remote IP : 30.149.115.231
Remote IP : 145.227.120.104
Remote IP : 45.113.215.205
Remote IP : 110.124.243.96
Remote IP : 110.124.243.96
IP 地址数: 8
```

图 3-25　例 3-28 测试代码运行结果

如文件 3-36 所示，第 9～20 行随机产生 10 个 IP 地址。为模拟同一 IP 地址多次访问，第 16～17 行将某次（本例为第 2 次）产生的 IP 地址做备份，随后将该备份地址作为第 9 次和第 10 次的 IP 地址（第 19、20 行）。第 22、23 行将当前系统时间作为 HLL 的键，第 24 行将 IP 地址保存到 HLL 中，执行去重计数。第 25 行设定记录在 Redis 中保存的时长。该测试代码的运行结果如图 3-25 所示。

3.9 Spring 操作 Redis 位图

位图是通过计算机存储数据的最小单位——位来表示某个元素对应的值或者状态的方法。一个位的值或者是 0，或者是 1，也就是说一个位只能表示两种状态。Redis 位图是字符串类型的扩展，可以将字符串视为位向量，进而对一个或多个字符串执行逐位操作。例如，字符串"big"，字母 b、i 和 g 对应的 ASCII 码分别为 0x62(01100010B)、0x69(01101001B)和 0x67(01100111B)，则字符串"big"在 Redis 的存储情况为：

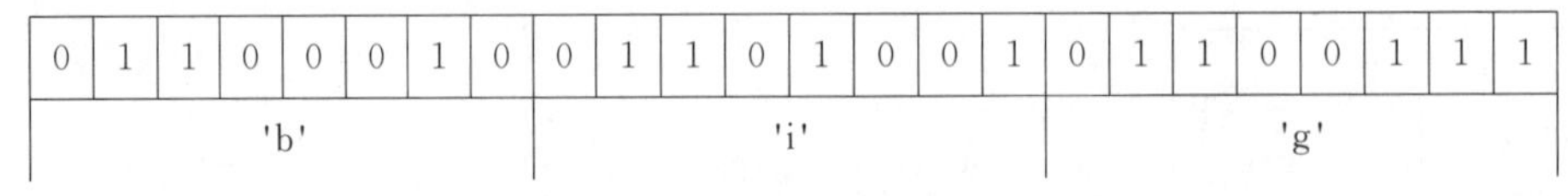
<table>
<tr><td>0</td><td>1</td><td>1</td><td>0</td><td>0</td><td>0</td><td>1</td><td>0</td><td>0</td><td>1</td><td>1</td><td>0</td><td>1</td><td>0</td><td>0</td><td>1</td><td>0</td><td>1</td><td>1</td><td>0</td><td>0</td><td>1</td><td>1</td><td>1</td></tr>
<tr><td colspan="8">'b'</td><td colspan="8">'i'</td><td colspan="8">'g'</td></tr>
</table>

由于位图是字符串类型的扩展，因此位图类型的数据也可以使用字符串类型的命令，主要是 SET 和 GET。可以通过 SETBIT 命令设置各位的值，来保存字符串。如，在 Redis 中执行下述命令：

```
redis > SETBIT bk 0 0
(integer) 0
redis > SETBIT bk 1 1
(integer) 0
redis > SETBIT bk 2 1
(integer) 0
redis > SETBIT bk 3 0
```

```
(integer) 0
redis > SETBIT bk 4 0
(integer) 0
redis > SETBIT bk 5 0
(integer) 0
redis > SETBIT bk 6 1
(integer) 0
redis > SETBIT bk 7 0
(integer) 0
redis > GET bk
"b"
```

SETBIT 命令的格式为：SETBIT key offset value。功能：针对 key 存储的字符串值，设置或清除指定偏移量 offset 上的位，位的设置或清除取决于 value 值，即 1 或 0。当 key 不存在时，会创建一个新的字符串。而且这个字符串的长度会伸展，直到可以满足指定的偏移量 offset($0\leqslant offset<2^{32}$)，在伸展过程中，新增的位的值被设置为 0。如果想要设置位图的非零初值，一种方式就是将每个位逐个设置为 0 或 1，但是这种方式比较麻烦；另一种方式是可以直接使用 SET 命令存储一个字符串。如准备设置一个 8 位长度的位图的初值，其中的第 2、3、4 位为 1，其余位为 0，即初值为 00111000，此初值为字符'8'的 ASCII 码。因此，可通过下述命令完成初值设置。

```
redis > set mk "8"
OK
redis > GETBIT mk 2
(integer) 1
redis > GETBIT mk 3
(integer) 1
redis > GETBIT mk 4
(integer) 1
redis > GETBIT mk 1
(integer) 0
```

GETBIT 命令的格式为：GETBIT key offset。功能：返回 key 对应的字符串在 offset 位置的位，当 offset 大于值的长度时，返回 0；当 key 不存在时，可以认为 value 为空字符串，此时 offset 肯定大于空字符串长度，返回 0。Redis 中位图相关的常用命令如表 3-5 所示。

表 3-5 Redis 中位图相关的常用命令

命　令	含　义
GETBIT key offset	返回以键(key)存储的字符串值中偏移量(offset)处的位值
SETBIT key key offset value	设置或清除键(key)处存储的字符串值中的偏移位。根据值(value 可以是 1 或 0)设置或清除位
BITCOUNT key [start end]	获取位图指定范围中位值为 1 的个数，如果不指定 start 与 end，则取所有
BITOP op destKey key1 [key2...]	执行多个 BitMap 的 AND(交集)、OR(并集)、NOT(非)、XOR(异或)操作并将结果保存在 destKey 中

位图基于位进行存储，所以具有节省空间、操作快速、方便扩容等优点。位图有很多应用场景，如：

(1) 记录用户在线状态，只需要一个键，将用户 id 作为偏移量。如果用户在线就设置为 1，不在线就设置为 0，3 亿用户只需要 36MB 的空间。命令示例如下：

```
$status = 1;
$redis->setBit('online', $uid, $status);
$redis->getBit('online', $uid);
```

（2）统计活跃用户数，使用时间作为缓存的键，用户 id 作为偏移量。如果当日活跃过就设置为 1，通过简单的计算就可得到用户在某时段的活跃情况。命令示例如下：

```
$status = 1;
$redis->setBit('active_20220708', $uid, $status);
$redis->setBit('active_20220709', $uid, $status);
$redis->bitOp('AND', 'active', 'active_20220708', 'active_20220709');
```

（3）用户签到。假设某网站有 1000 万用户，平均每人每年签到次数为 10 次，如果用关系数据库保存签到数据的话，1 年将产生 1 亿条签到数据。这样就需要保存和处理大量的意义不大的数据。如果将用户每日签到信息保存到 Redis 中，则以位图形式处理就会节约很多空间。如果用户签到以 1 表示，未签到则以 0 表示。在执行按月统计用户签到信息时，只需要 32 位数据，每个用户 4B 即可保存。1000 万用户一年的签到数据仅仅 480MB。如，12 月前 10 天某用户的签到情况如下：除 12 月 3 日、4 日、7 日和 9 日没有签到，其余已签到。签到数据可以用位图表示如下：

1	1	0	0	1	1	0	1	0	1
12-1	12-2	12-3	12-4	12-5	12-6	12-7	12-8	12-9	12-10

要统计该用户在 12 月前 10 天签到的总天数，只要计算位图结构中 1 的数量即可。处理用户签到程序的代码如文件 3-37 所示。

【文件 3-37】 TestBitmapOperations.java

```
1     @RunWith(SpringJUnit4ClassRunner.class)
2     @ContextConfiguration(classes = RedisTemplateConfig.class)
3     public class TestBitmapOperations {
4         @Autowired
5         private StringRedisTemplate template;
6
7         @Test
8         public void testAttendance() throws NoSuchAlgorithmException {
9             String uid = "1001";
10            LocalDateTime now = LocalDateTime.now();
11            String date =
12                now.format(DateTimeFormatter.ofPattern("yyyyMM"));
13            String key = uid + ":" + date;
14            SecureRandom random = SecureRandom.getInstance("SHA1PRNG");
15            random.setSeed(2L);
16            for(int i = 0;i < 31;i++) {
17                int a = random.nextInt(2);
18                template.opsForValue().setBit(key, i, a == 1);
19                System.out.print(a + " ");
20            }
21            List<Long> result = template.opsForValue().bitField(key,
22              BitFieldSubCommands.create().get(BitFieldSubCommands
23               .BitFieldType.unsigned(31)).valueAt(0));
```

```
24            Long num = result.get(0);
25            int count = 0;
26            while (num > 0) {
27            //让这个数字与 1 做与运算,得到数字的最后一个位判断这个数字是否为 0
28                if ((num & 1) == 1)
29                    //如果为 1,则表示签到 1 次
30                    count++;
31                num >>>= 1;
32            }
33            System.out.println("num = " + count);
34        }
35    }
```

如文件 3-37 所示,第 2 行指定了元数据的配置类为 RedisTemplateConfig. class(代码可参考文件 3-6)。第 9～13 行设定存入 Redis 中的键的格式为“用户 id:年月”。第 14、15 行随机产生 1、0 两个数字,用来模拟用户签到数据。第 16～20 行将模拟的签到数据(按每月 31 天计算)存入 Redis 的位图中,同时将签到数据输出到控制台,以便核对。第 21～34 行进行当月签到次数的统计,即计算位图中数字 1 出现的次数。其中,第 22、23 行获取本月的全部签到数据,其返回结果是一组十进制数字。随后,用这个数字的每一位与数字 1 做与运算,根据与运算的结果判定当天是否签到(第 27～33 行)。最后,经过与运算得到的 1 的个数即为签到的天数。程序运行结果如图 3-26 所示。

```
0 1 0 0 1 0 1 1 0 1 1 0 0 0 0 0 0 1 1 1 1 1 1 1 0 1 0 1 1 0 0 签到天数 = 16
```

图 3-26 签到统计程序运行结果

3.10 键绑定操作子接口

前面几节分别介绍了 ValueOperations、ListOperations 等操作接口。这些操作接口有一个共性,在进行某个具体操作时,都需要指定键,例如,使用这些操作接口添加值时,如下:

```
1    SetOperations -> add(K key, V... values)
2    ZSetOperations -> add(K key, Set < ZSetOperations.TypedTuple < V >> tuples)
3    ValueOperations -> set(K key, V value)
4    ListOperations -> leftPush(K key, V value)
```

其实在整个操作过程中,可能并没有更换键,只是反复对一个键进行设置、取值、删除操作。于是,RedisTemplate 类提供了一套便捷操作子接口——键绑定操作子接口,这些子接口提前将某个键和操作接口进行绑定。这样,在使用键绑定操作子接口进行操作时,就不需要传递键了。RedisTemplate 类提供的键绑定子接口如表 3-6 所示。

表 3-6 RedisTemplate 类提供的键绑定子接口

子 接 口	说 明
BoundGeoOperations	地理空间数据键绑定操作
BoundHashOperations	哈希键绑定操作
BoundKeyOperations	键绑定操作

续表

子　接　口	说　　明
BoundListOperations	列表键绑定操作
BoundSetOperations	集合键绑定操作
BoundValueOperations	字符串(或值)键绑定操作
BoundZSetOperations	有序集合键绑定操作

下面通过一些示例演示怎样利用键绑定操作子接口操作哈希、列表、集合、字符串、有序集合数据类型。

【例 3-29】 利用键绑定子接口操作键为 bound-value 的字符串类型(String)的数据。分别执行添加、修改和追加值操作。测试代码如文件 3-38 所示。

【文件 3-38】 例 3-29 测试代码

```
1    @Test
2    public void testValueBoundOperations() throws InterruptedException{
3      BoundValueOperations<String,String> ops =
4        template.boundValueOps("bound-value");
5      //添加值
6      ops.set("value1");
7      System.out.println(ops.get());
8      //修改值
9      ops.set("value2");
10     System.out.println(ops.get());
11     //追加值
12     ops.append(" append");
13     System.out.println(ops.get());
14     ops.set("value3", Duration.ofMillis(3000));
15     System.out.println(ops.get());
16     Thread.sleep(5000);
17     System.out.println(ops.get());
18   }
```

```
value1
value2
value2 append
value3
null
```

图 3-27　例 3-29 测试代码运行结果

如文件 3-38 所示，第 14 行在设置值 value3 的同时，指定了该值在 Redis 中的缓存时长。第 16 行设置当前程序挂起 5 秒，等待键 bound-value 对应的值 value3 过期，第 17 行再次获取键 bound-value 对应的值。运行此测试代码，运行结果如图 3-27 所示。

【例 3-30】 利用键绑定子接口操作键为 bound-list 的列表类型(List)的数据。分别执行添加、修改和删除操作。测试代码如文件 3-39 所示。

【文件 3-39】 例 3-30 测试代码

```
1    @Test
2    public void testListBoundOperations() {
3       BoundListOperations<String, String> ops =
4          template.boundListOps("bound-list");
5       //添加
6       ops.leftPush("value1");
```

```
7       ops.rightPush("value2");
8       ops.rightPushAll("value3","value4");
9       System.out.println(ops.range(0, -1));
10      //修改
11      ops.set(0,"new value");
12      System.out.println(ops.range(0, -1));
13      //删除
14      ops.remove(1,"value3");
15      System.out.println(ops.range(0, -1));
16  }
```

运行此测试代码,运行结果如图 3-28 所示。

```
[value1, value2, value3, value4]
[new value, value2, value3, value4]
[new value, value2, value4]
```

图 3-28 例 3-30 测试代码运行结果

【例 3-31】 利用键绑定子接口操作键为 bound-hash 的哈希类型的数据。分别执行添加、修改和扫描操作。测试代码如文件 3-40 所示。

【文件 3-40】 例 3-31 测试代码

```
1   @Test
2   public void testHashBoundOperations() {
3     BoundHashOperations<String, String,String> ops =
4        template.boundHashOps("bound-hash");
5     //添加
6     ops.put("k1","v1");
7     Map<String,String> map = new HashMap<>();
8     map.put("k2","v2");  map.put("k3","100");
9     ops.putAll(map);
10    System.out.println(ops.entries()); System.out.println("--------");
11    //修改
12    ops.increment("k3",10);
13    System.out.println("k3 = " + ops.get("k3"));
14    ops.delete("k1");
15    System.out.println(ops.entries()); System.out.println("--------");
16    //扫描
17    Cursor<Map.Entry<String,String>> cursor =
18      ops.scan(ScanOptions.NONE);
19    cursor.forEachRemaining(entry -> System.out.println(
20      entry.getKey() + " = " + entry.getValue()));
21  }
```

运行此测试代码,运行结果如图 3-29 所示。

```
{k1=v1, k3=100, k2=v2}
--------
k3 = 110
{k3=110, k2=v2}
--------
k2=v2
k3=110
```

图 3-29 例 3-31 测试代码运行结果

【例 3-32】 利用键绑定子接口操作键为 bound-set 的集合类型的数据。分别执行添加、移除和扫描操作。测试代码如文件 3-41 所示。

【文件 3-41】 例 3-32 测试代码

```
1     @Test
2     public void testSetBoundOperations() {
3         BoundSetOperations<String,String> ops =
4           template.boundSetOps("bound-set");
5         ops.add("v1","v2","v3","v4","v5");
6         System.out.println(ops.members());
7         System.out.println("--------");
8         ops.remove("v3");
9         Cursor<String> cursor = ops.scan(ScanOptions.NONE);
10        if(cursor != null)
11            cursor.forEachRemaining(e -> System.out.print(e+" "));
12    }
```

运行此测试代码，运行结果如图 3-30 所示。

```
[v5, v1, v2, v3, v4]
--------
v5 v1 v2 v4
```

图 3-30 例 3-32 测试代码运行结果

【例 3-33】 利用键绑定子接口操作键为 bound-zset 的有序集合类型的数据。分别执行添加、移除和扫描操作。测试代码如文件 3-42 所示。

【文件 3-42】 例 3-33 测试代码

```
1     @Test
2     public void testZSetBoundOperations() {
3         BoundZSetOperations<String, String> ops =
4             template.boundZSetOps("bound-zset");
5         ops.add("GuanYu",32.0D);
6         ops.add("LiuBei",21.9D);
7         ops.add("ZhangFei",41.0D);
8         ops.add("CaoCao",100.0D);
9         System.out.println(ops.range(0, -1));
10        ops.remove("CaoCao");
11        System.out.println("--------");
12        Cursor<ZSetOperations.TypedTuple<String>> tups  =
13            ops.scan(ScanOptions.NONE);
14        tups.forEachRemaining(tup -> System.out.println(
15            tup.getValue() + " = " + tup.getScore()));
16    }
```

运行此测试代码，运行结果如图 3-31 所示。

```
[LiuBei, GuanYu, ZhangFei, CaoCao]
--------
LiuBei=21.9
GuanYu=32.0
ZhangFei=41.0
```

图 3-31 例 3-33 测试代码运行结果

在电商或社交网络中，经常有自动补全的需求。用户在文本框输入的文字时，会自动弹出一个下拉框，给用户提供候选词，方便输入。例如，在百度主页输入“大学”两个字，会有相应的候选词出现，如图 3-32 所示。

关于自动补全这个功能，可以使用 Redis 的有序集合实现。有序集合会默认将存入的字符串按分数进行升序排列。如果存入的字符串的分数相等，则按照字符串中字母的字典序升序排列。如存入分数相同的三个字符串"aab"、"abb"、"aaba"，则这三个字符串在有序集合中的顺序为"aab"、"aaba"、"abb"。这样，就可以把提供给用户的候选词存放到有序集合中，并且保持这些词的分数相同。那么，要查找带有前缀为 abc 的词，就是查找前缀介于 abbz 和 abd 之间的词，如图 3-33 所示，即锁定图中 p 和 q 的位置。

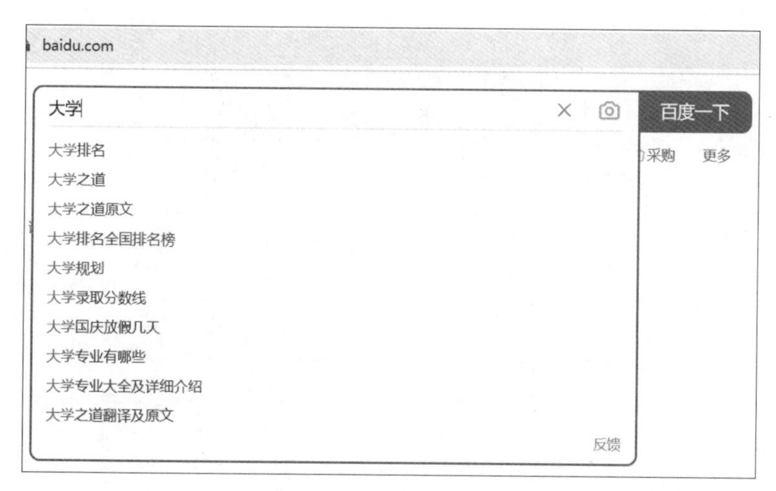

图 3-32 百度主页提供的候选词

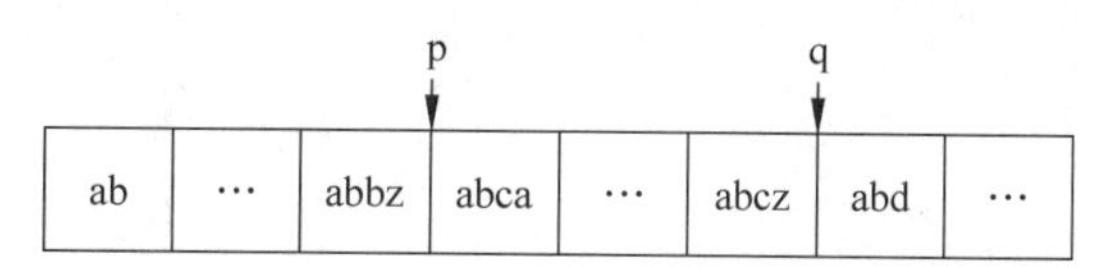

图 3-33 有序集合中存放的候选词

要锁定 p 和 q 的位置,以查找前缀为 abc 的词(不包括 abc 本身)为例,只需要向有序集合中插入 abc`和 abc{两个元素。因为在 ASCII 编码中,排在字符 a 前面的字符是反引号(`)字符,排在字符 z 后面的是左花括号({)字符。即 abc`会排在所有拥有 abca 前缀的词的前面,而位于所有拥有 abbz 前缀的词的后面。同理,abc{会位于所有拥有 abcz 前缀的词的后面,而位于所有拥有 abd 前缀的词的前面。两个元素进入有序集合后,就会占据图 3-33 中 p 和 q 指向的位置,如图 3-34 所示。

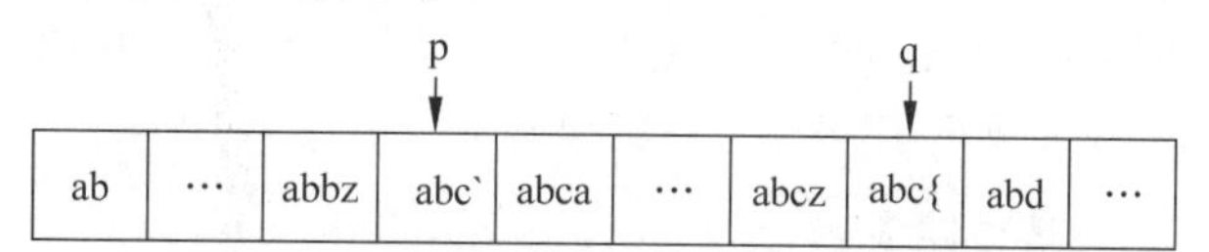

图 3-34 有序集合中插入两个元素后的情况

随后,通过 p 和 q,即 abc`和 abc{在有序集合中的位置便可确定提供给用户的候选词集合。

【例 3-34】 准备一组候选词,根据用户输入的单词前缀给出最多 10 个候选词,实现自动补全功能。测试代码如文件 3-43 所示。

【文件 3-43】 AutoCompletion.java

```
1   public class AutoCompletion {
2       public Set<String> doAutoCompletion(StringRedisTemplate template,
3           String prefix){
4           String[] names = {
5                   "astra","astrid","astrix","athena","athene","atlanta",
6                   "barb","barbara","barbe","barbey","barrie","barry",
7                   "caden","cadesa","cafesse","cagey","caril","carine"
```

```
8           };
9           BoundZSetOperations<String, String> ops =
10              template.boundZSetOps("words");
11          for(int i = 0;i<names.length;i++)
12              ops.add(names[i],0);
13          String p = prefix + "~";
14          String q = prefix + "{";
15          ops.add(p,0);
16          ops.add(q,0);
17          Long begin_index = ops.rank(p);
18          Long end_index = ops.rank(q);
19          if(end_index - begin_index > 10)
20              end_index = begin_index + 11;
21          Set<String> tips = ops.range(begin_index + 1,end_index - 1);
22          ops.remove(p);
23          ops.remove(q);
24          return tips;
25      }
26  }
```

如文件 3-43 所示，第 4～8 行将一组候选词存放到 names 数组中。第 9、10 行绑定操作有序集合的键 words。第 11、12 行将 names 数组中存放的候选词加入有序集合，同时设定所有元素的分数为 0。这就意味着所有加入有序集合的候选词会按照字母表顺序升序排列。第 13、14 行根据用户输入的词语前缀生成两个定位符 p 和 q。第 15、16 行将 p、q 加入有序集合。第 17、18 行获取两个定位符在有序集合中的下标（位置），这样，在两个定位符之间的元素组成了返回给用户的候选词集合。第 21 行从有序集合中取出一定数量的候选词，随后删除两个定位符（第 22、23 行）。

根据文件 3-43 中给出的候选词，编写测试代码，从候选词中查找前缀为 ast 的单词作为提供给用户的候选词（最多取出 10 个）。测试代码如文件 3-44 所示。

【文件 3-44】 TestAutoCompletion.java

```
1   @RunWith(SpringJUnit4ClassRunner.class)
2   @ContextConfiguration(classes = RedisTemplateConfig.class)
3   public class TestAutoCompletion {
4       @Autowired
5       private StringRedisTemplate template;
6
7       @Test
8       public void testAutoCompletion(){
9           AutoCompletion app = new AutoCompletion();
10          Set<String> tips = app.doAutoCompletion(template,"ast");
11          tips.forEach(System.out::println);
12      }
13  }
```

运行测试代码，从文件 3-43 提供的候选词中选择出含有 ast 前缀的词，测试结果如图 3-35 所示。

```
✔ Tests passed: 1 of 1 test – 157 ms
/Library/Java/JavaVirtualMac
SLF4J: Failed to load class
SLF4J: Defaulting to no-oper
SLF4J: See http://www.slf4j.
astra
astrid
astrix
```

图 3-35　自动补全功能测试结果

3.11　RedisTemplate 类的通用方法

RedisTemplate 类除了提供 Redis 常用数据类型的操作接口外，还提供了操作缓存的方法，如删除键、判断键是否存在、键设置过期时间、键移动等方法。本节将介绍这些方法。

(1) 方法原型：List < RedisClientInfo > getClientList()；功能：获取已连接的客户端的信息。

(2) 方法原型：Boolean expire(K key,long timeout,TimeUnit unit)；功能：设置指定键 key 的生存时间。

(3) 方法原型：Boolean persist(K key)；功能：移除指定键 key 的过期时间；对应的 Redis 命令：PERSIST。

(4) 方法原型：void rename(K oldKey,K newKey)；功能：重命名键；对应的 Redis 命令：RENAME。

(5) 方法原型：Boolean hasKey(K key)；功能：判断键 key 是否存在；对应的 Redis 命令：EXISTS。

(6) 方法原型：Boolean delete(K key)；重载形式：Long delete(Collection < K > keys)；功能：删除指定的键 key(s)；对应的 Redis 命令：DEL。

【例 3-35】 应用上述方法完成以下要求：①获取客户端的信息；②将键 expire-key 的过期时间设置为 3 秒；③移除 expire-key 的过期时间；④将键 no-key 重命名为 has-key，并判断 has-key 和 no-key 是否存在；⑤删除键 delete-key。测试代码如文件 3-45 所示。

【文件 3-45】　例 3-35 测试代码

```
1    @Test
2    public void testCommonOperations1() {
3        //①获取客户端连接信息
4        System.out.println("---- exercise 1 ----");
5        List < RedisClientInfo > list = template.getClientList();
6        if(list != null)
7            list.forEach(System.out::println);
8        //②设置过期时间
9        System.out.println("---- exercise 2 ----");
10       template.opsForValue().set("expire - key","data");
11       template.expire("expire - key", Duration.ofMillis(3000));
12       System.out.println("expire - key exprired within "
13           + template.getExpire("expire - key") + " ms");
```

```
14        //③清理过期时间
15        System.out.println("---- exercise 3 ----");
16        template.persist("expire-key");
17        System.out.println("expire=" + template.getExpire("expire-key"));
18        //④重命名键
19        System.out.println("---- exercise 4 ----");
20        template.opsForValue().set("no-key","value");
21        template.rename("no-key","has-key");
22        //⑤判断键是否存在
23        System.out.println("---- exercise 5 ----");
24        template.opsForValue().set("has-key","data");
25        Boolean flag = template.hasKey("has-key");
26        if(flag != null && flag)
27            System.out.println("键 has-key 存在");
28        else
29            System.out.println("键 has-key 不存在");
30        System.out.println(template.hasKey("no-key")?"键 no-key 存在":
31            "键 no-key 不存在");
32        //⑥删除键
33        System.out.println("---- exercise 6 ----");
34        template.opsForValue().set("delete-key","value");
35        System.out.print("删除前: ");
36        System.out.println(template.hasKey("delete-key")?
37            "delete-key 存在":"delete-key 不存在");
38        template.delete("delete-key");
39        System.out.print("删除后: ");
40        System.out.println(template.hasKey("delete-key")?
41            "delete-key 存在":"delete-key 不存在");
42    }
```

如文件 3-45 所示,本例使用的 template 是 RedisTemplate<String,String>的实例,其配置代码可参考文件 3-6。运行此测试代码,控制台的输出如图 3-36 所示。

```
---- exercise 1 ----
{sub=0, flags=N, multi=-1, qbuf=0, id=211, addr=127.0.0.1:51841,
 events=r, psub=0, idle=7, qbuf-free=0, oll=0, omem=0, name=, obl=0,
 cmd=command, fd=14, age=7, db=0}
{sub=0, flags=N, multi=-1, qbuf=26, id=212, addr=127.0.0.1:51848,
 events=r, psub=0, idle=0, qbuf-free=32742, oll=0, omem=0, name=,
 obl=0, cmd=client, fd=11, age=0, db=0}
---- exercise 2 ----
expire-key exprired within 3 ms
---- exercise 3 ----
expire = -1
---- exercise 4 ----
键has-key存在
键no-key不存在
---- exercise 5 ----
删除前: delete-key存在
删除后: delete-key不存在
```

图 3-36 例 3-35 测试代码运行后控制台的输出

(7) 方法原型: Set<K>keys(K pattern); 功能: 查找与给定模式匹配的所有键; 对应的 Redis 命令: KEYS。

其中,关于参数 pattern 的定义有以下几种情况:

① 问号(?)匹配符：表示仅匹配一个字母，如 h?llo 可匹配 hello、hallo 和 hxllo 等。

② 星号(*)匹配符：表示匹配 0 个或多个字母，如 h*llo 可匹配 hllo 和 heeeello 等。

③ 列表([])匹配：表示仅匹配列表中的一个字符，如 h[ae]llo 可匹配 hello 和 hallo，但不能匹配 hillo。也可以在列表中指定一个有序序列的首尾字符，则可以匹配包括首尾字符在内的一个字符，如 h[a-b]llo 可匹配 hallo 和 hbllo。此外，可以用“^”符号结合列表符号“[”和“]”实现排除某个字符，如 h[^e]llo 可匹配 hallo、hbllo 等，但不能匹配 hello。

此外，方法原型定义中的返回值类型 Set 中的泛型 K 与 RedisTemplate<K,V>中定义的泛型 K 类型一致。在本节中，已将泛型 K 定义为 String。

【例 3-36】 将<CaoCao,v1>、<CaoPi,v21>、<CaoZhang,v22>和<CaoXiong,v23>分别加入 Redis。要求：①找到 Cao 开头的所有键及其值；②找到 Cao 开头的跟随两个字母的键及其值；③找到 Cao 开头的随后为 P 或 C 的键及其值；④找到以 Cao 开头的随后为 X 或 Z，并以 g 为结尾的键及其值；⑤找到以 Cao 开头的随后不含字母 C 的键及其值。测试代码如文件 3-46 所示。

【文件 3-46】 例 3-36 测试代码

```
1   @Test
2   public void testCommonOperations2() {
3       template.opsForValue().set("CaoCao","v1");
4       template.opsForValue().set("CaoPi","v21");
5       template.opsForValue().set("CaoZhang","v22");
6       template.opsForValue().set("CaoXiong","v23");
7       //1-1 pattern 1 "*"
8       template.keys("Cao*").iterator().forEachRemaining(e->
9           System.out.println(e+" : "+template.opsForValue().get(e)));
10      //1-2 pattern 2 "?"
11      template.keys("Cao??").iterator().forEachRemaining(e->
12          System.out.println(e+" : "+template.opsForValue().get(e)));
13      //1-3 pattern 3 "[PC]*"
14      template.keys("Cao[PC]*").iterator().forEachRemaining(e->
15          System.out.println(e+" : "+template.opsForValue().get(e)));
16      //1-4 pattern 4 "Cao[X-Z]???g" or "Cao[X-Z]*"
17      template.keys("Cao[X-Z]*").iterator().forEachRemaining(e->
18          System.out.println(e+" : "+template.opsForValue().get(e)));
19      //1-5 pattern 5 "Cao[^C]*"
20      template.keys("Cao[^C]*").iterator().forEachRemaining(e->
21          System.out.println(e+" : "+template.opsForValue().get(e)));
22  }
```

运行此测试代码，控制台的输出如图 3-37 所示。

(8) 方法原型：K randomKey()；功能：从键空间中随机返回一个键；对应的 Redis 命令：RANDOMKEY。

(9) 方法原型：DataType type(K key)；功能：确定键存储的数据的类型；对应的 Redis 命令：TYPE。

【例 3-37】 向 key0～key7 共 8 个键中存入 8 个随机选择的字符串，再从此 8 个键中随机取出 3 个键，并判断对应的值的类型。测试代码如文件 3-47 所示。

```
CaoXiong : v23
CaoPi : v21
CaoCao : v1
CaoZhang : v22
--------
CaoPi : v21
--------
CaoPi : v21
CaoCao : v1
--------
CaoXiong : v23
CaoZhang : v22
--------
CaoXiong : v23
CaoPi : v21
CaoZhang : v22
```

图 3-37 例 3-36 测试代码运行后控制台的输出

【文件 3-47】 例 3-37 测试代码

```
1    @Test
2    public void testCommonOperations3() {
3        for(int i = 0;i < 8;i++)
4            template.opsForValue().set(
5                "key" + i,UUID.randomUUID().toString());
6        for(int j = 0;j < 3;j++){
7            String rk = template.randomKey();
8            System.out.println(rk + ":" + template.opsForValue().get(rk));
9            System.out.println("type is:" + template.type(rk).name());
10       }
11   }
```

运行此测试代码，控制台的输出如图 3-38 所示。

```
key5:62277d05-f24f-4a1a-b065-5db948043859
type is : STRING
key6:64b964e9-138c-4503-9cac-77b3d6874c24
type is : STRING
key0:9e760887-1f96-4a83-a0c2-6a03debd9ad5
type is : STRING
```

图 3-38 例 3-37 测试代码运行后控制台的输出

（10）方法原型：Boolean move(K key,int dbIndex)；功能：将键 key 移动到由参数 dbIndex 指定的数据库；对应的 Redis 命令：MOVE；Redis 默认配置中共有 16 个数据库，索引为 0～15。在 Redis 命令提示符下可以用

```
select dbIndex
```

命令切换不同的数据库。不同数据库之间的数据没有任何关联，甚至可以存在相同的键。

（11）方法原型：byte[]dump(K key)；功能：执行数据备份；对应的 Redis 命令：DUMP。该方法可执行 Redis 的转储命令并返回结果。Redis 使用非标准的序列化机制并包含校验和信息，因此该方法返回原始字节，而没有使用 ValueSerializer 进行反序列化。可使用转储的返回值(byte[])作为要备份的值的参数。

（12）方法原型：public void restore(K key,byte[] value,long timeToLive,TimeUnit unit,boolean replace)；功能：执行 Redis 还原命令。其中，参数 key 指定要还原的键，即目

标键；参数 value 指定转储对象返回的要还原的值；参数 timeToLive 指定目标键的到期时间，0 表示无到期时间；参数 replace 指定为 true 表示替换可能存在的值，默认值 false。对应的 Redis 命令：RESTORE。

【例 3-38】 完成如下两个要求：①将 move-key 键移动到索引为 4 的数据库；②将 backup-key 键备份到 backup-key-bak 键，并设置键 backup-key-bak 的生存时间为 30 分钟。测试代码如文件 3-48 所示。

【文件 3-48】 例 3-38 测试代码

```
1    @Test
2    public void testCommonOperations4() {
3      template.opsForValue().set("move-key","data");
4      System.out.println("移动前" + template.opsForValue()
5       .get("move-key"));
6      template.move("move-key",4);
7      System.out.println("移动后" + template.opsForValue()
8        .get("move-key"));
9      template.opsForValue().set("backup-key","this is important");
10     byte[] backup = template.dump("backup-key");
11     if(backup != null) {
12       template.restore("backup-key-bak", backup, 30,
13            TimeUnit.MINUTES);
14       System.out.println("backup data: " +
15            template.opsForValue().get("backup-key-bak"));
16       }
17   }
```

执行移动操作后（第 6 行），move-key 从 0 号数据库被移动到 4 号数据库。当再次执行获取值操作时（第 7、8 行），返回 null。此时，在 Redis 客户端执行 select 4 命令，可以在 4 号数据库找到被移动过来的 move-key 键：

```
redis> select 4
OK
redis[4]> keys *
1) "move-key"
```

运行此测试代码，控制台的输出如图 3-39 所示。

```
移动前data
移动后null
backup data: this is important
```

图 3-39 例 3-38 测试代码运行后控制台的输出

（13）方法原型：List<V> sort(SortQuery<K> query)；功能：对要查询的元素进行排序；对应的 Redis 命令：SORT。

【例 3-39】 对给定的 8 位散文家（SuXun，HanYu，Wang ' anShi，SuShi，LiuZongYuan，SuZhe，OuYangXiu，ZengGong）按名字的字母序升序排列，输出排序后的前三位。测试代码如文件 3-49 所示。

【文件 3-49】 例 3-39 测试代码

```
1    @Test
2    public void testCommonOperations3() {
3       String sortKey = "sortKey";
4       String[] names = {"SuXun","HanYu","Wang'anShi","SuShi",
5          "LiuZongYuan","SuZhe","OuYangXiu","ZengGong"};
```

```
6        String[] marks = {"1009","768","1021","1037",
7           "773","1039","1007","1019"};
8        template.delete(sortKey);
9        if (!template.hasKey(sortKey)) {
10          for (int i = 0; i < 8; i++) {
11             template.boundSetOps(sortKey).add(String.valueOf(i));
12             String hashKey = "hash" + i,
13                pid = String.valueOf(i),
14                pname = names[i],
15                pmark = marks[i];
16             template.boundHashOps(hashKey).put("id", pid);
17             template.boundHashOps(hashKey).put("name", pname);
18             template.boundHashOps(hashKey).put("mark", pmark);
19             System.out.printf("%s:{\"_id\":%s,\"Name\":
20                 %s,\"Mark\",%s}\n",hashKey, pid, pname, pmark);
21          }
22       }
23       SortQuery<String> sortQuery = SortQueryBuilder.sort(sortKey)
24          .by("hash*->name")
25          .alphabetical(true)
26          .limit(new SortParameters.Range(0,3))
27          .order(SortParameters.Order.ASC)
28          .get("hash*->id")
29          .get("hash*->name")
30          .get("hash*->mark").build();
31       System.out.println("----  ----");
32       List<String> list = template.sort(sortQuery);
33       for(int j = 0;j < list.size();j += 3)
34          System.out.printf("{\"ID\": %s, \"Name\": %s, \"Mark\", %s}\n",
35                list.get(j), list.get(j+1), list.get(j+2));
36    }
```

排序运行原理可以了解 Redis 的 SORT 命令（见 1.4.1 节）。运行此测试代码，控制台的输出如图 3-40 所示。

```
hash0 : {"_id": 0, "Name": SuXun, "Mark", 1009}
hash1 : {"_id": 1, "Name": HanYu, "Mark", 768}
hash2 : {"_id": 2, "Name": Wang'anShi, "Mark", 1021}
hash3 : {"_id": 3, "Name": SuShi, "Mark", 1037}
hash4 : {"_id": 4, "Name": LiuZongYuan, "Mark", 773}
hash5 : {"_id": 5, "Name": SuZhe, "Mark", 1039}
hash6 : {"_id": 6, "Name": OuYangXiu, "Mark", 1007}
hash7 : {"_id": 7, "Name": ZengGong, "Mark", 1019}
---- ----
{"ID": 1, "Name": HanYu, "Mark", 768}
{"ID": 4, "Name": LiuZongYuan, "Mark", 773}
{"ID": 6, "Name": OuYangXiu, "Mark", 1007}
```

图 3-40　例 3-39 测试代码运行后控制台的输出

在 RedisTemplate 中，定义了几个 execute()方法，这些方法是 RedisTemplate 的核心方法。RedisTemplate 中很多其他方法均是通过调用 execute()方法来执行具体的操作。表 3-7 列举了 execute()方法的 6 种重载形式。

表 3-7 RedisTemplate 定义的 execute()方法

方法原型	说明
<T> T execute(RedisCallback<T> action)	在 Redis 连接中执行给定的操作
<T> T execute(RedisCallback<T> action, boolean exposeConnection)	在 Redis 连接中执行给定的操作,参数 exposeConnection 表示是否要暴露当前连接,如果为 true,那么就可以在回调函数中使用当前连接对象
<T> T execute(RedisCallback<T> action, boolean exposeConnection, boolean pipeline)	在连接中执行给定的操作,参数 exposeConnection 的含义同上,参数 pipeline 表示是否开启流水线(见 5.3 节)
<T> T execute(RedisScript<T> script, List<K> keys, Object... args)	执行给定的 Redis 脚本(Redis Script)
<T> T execute(RedisScript<T> script, RedisSerializer<?> argsSerializer, RedisSerializer<T> resultSerializer, List<K> keys, Object... args)	执行给定的 Redis 脚本,使用提供的 RedisSerializer 序列化脚本参数和结果
<T> T execute(SessionCallback<T> session)	执行 Redis 会话

使用 RedisTemplate 直接调用 opsFor**()方法来操作 Redis 时,每执行一条命令时都要重新获取一个连接,因此很耗资源。可以调用 execute()方法,让一个连接直接执行多次 Redis 操作语句。

【例 3-40】 利用一个 Redis 连接完成字符串数据的存取操作。测试代码如文件 3-50 所示。

【文件 3-50】 TestExecuteMethod.java

```
1   @RunWith(SpringJUnit4ClassRunner.class)
2   @ContextConfiguration(classes = ExecuteMethodConfig.class)
3   public class TestExecuteMethod {
4       @Autowired
5       private RedisTemplate<String,String> template;
6       @Test
7       public void testExecuteByRedisCallback(){
8           template.execute((RedisCallback<Object>) connection -> {
9               connection.stringCommands().set("key".getBytes(),
10                      "hello,redis".getBytes());
11              byte[] res = connection.stringCommands()
12                      .get("key".getBytes());
13              System.out.println(new String(res));
14              return null;
15          });
16      }
17  }
```

如文件 3-50 所示,第 8 行指定传递给 execute()方法的参数为 RedisCallback(接口)类型的对象。而 RedisCallback 接口中只定义了一个 doInRedis(RedisConnection connection)方法。因此,第 9~12 行可调用 RedisConnection 接口中定义或继承的方法。RedisConnection 接口代表了一个与 Redis 服务器的连接,它是各种 Redis 客户端(或驱动程序)的抽象。并且,RedisConnection 接口继承了 RedisStringCommands、RedisKeyCommands 等众多 Redis 操作接口,可以编程形式完成绝大部分的 Redis 操作。第 9 行调用 RedisConnection 接口的

stringCommands()方法获取 RedisStringCommands 接口类型对象，并调用 RedisStringCommands 接口的 Boolean set(byte[]key,byte[]value)和 byte[]get(byte[]key)方法完成字符串存取操作。运行此代码，控制台输出 hello,redis 字符串。

【例 3-41】 利用一个 Redis 会话完成字符串数据的存取操作。测试代码如文件 3-51 所示。

【文件 3-51】 例 3-41 测试代码

```
1   @Test
2   public void testExecuteBySessionCallback(){
3       template.execute(new SessionCallback<Object>(){
4           public String execute(RedisOperations operations)
5               throws DataAccessException {
6               BoundValueOperations<String, String> bops =
7                   operations.boundValueOps("key3");
8               bops.set("hello,world");
9               System.out.println(bops.get());
10              return null;
11          }
12      });
13  }
```

运行此测试代码，控制台输出 hello,world 字符串。用于执行 Redis 脚本的 execute()方法的案例见 5.4.3 节。

3.12 序列化和反序列化

虽然 Redis 支持各种数据类型，但 Redis 中存储的数据只有字节。因此，要利用 Redis 存储 Java 程序中的对象，就要将对象转换为字节数组或字符串再保存到 Redis 中。将对象转换为可传输(或可存储)的字节序列或字符串的过程称为序列化；将字节序列或字符串还原为对象的过程称为反序列化。进行序列化就是为了对象能够通过网络传输和跨平台存储。不同的计算机系统能够识别和处理的数据的通用格式是二进制数据(或纯文本数据)。因此，需要将对象按照一定规则转换为字节数组或字符串才能实现跨平台传输和存储的目的，这就是序列化。当需要对象时，再按这个规则把对象还原出来，这就是反序列化。作为键值型数据库，Redis 写入数据时，可以分别指定键和值的序列化机制。此外，还可以使用 Redis 哈希类型来实现更复杂的结构化对象映射，Spring Data Redis 提供了将 Java 对象映射到哈希的各种策略。本节将分别介绍键值序列化机制和对象-哈希序列化机制。

3.12.1 内置序列化器

将 Java 对象存储到 Redis 中时，需要进行序列化操作，如将 Java 对象序列化为 JSON 字符串。此时，Redis 中保存的内容为序列化后的 JSON 字符串。同样，如果要将序列化后的 JSON 字符串从 Redis 中取出，再转换为存储前的 Java 对象则需要反序列化操作，将 JSON 字符串转换为 Java 对象。在 Spring Data Redis 中，Java 对象和二进制数据之间的转换(反之亦然)可利用 org. springframework. data. redis. serializer 包(以下简称为 serializer

包）中提供的接口或类处理。

1. 序列化器

该包提供了两种类型的序列化器。

（1）基于 RedisSerializer 接口的双向序列化器。

（2）使用 RedisElementReader 接口和 RedisElementWriter 接口的元素读写器。

以上两种序列化器的主要区别是 RedisSerializer 接口主要序列化为字节数组（byte[]），而读写器使用字节缓冲区（ByteBuffer）。

2. RedisSerializer 接口的实现类

在 Spring Data Redis 中，serializer 包提供了以下几个 RedisSerializer 接口的实现类。

（1）JdkSerializationRedisSerializer 类：使用 JDK 的序列化器，将对象通过 ByteArrayOutputStream 类和 ByteArrayInputStream 类进行序列化和反序列化，最终 Redis 中将存储字节序列。该类是 RedisCache 类和 RedisTemplate 类默认的序列化器。限制：被序列化的类需要实现 Serializable 接口，而且 Redis 中存储的数据很不直观，序列化后的内容为十六进制数字或乱码。

（2）StringRedisSerializer 类：在键或值为字符串时，该类根据指定的字符集将字符串转换为字节序列（byte[]），也可以执行字节序列到字符串的反序列化。该类通过 String 类的 String. getBytes(Charset charset)和 String(byte[]byte，Charset charset)方法实现序列化和反序列化，是最轻量级和最高效的序列化器。

（3）Jackson2JsonRedisSerializer 类：使用 Jackson 和 Jackson Databind ObjectMapper 读取和写入 JSON 的序列化器。该序列化器可用于绑定到类型化的 Bean 或非类型化的 HashMap 实例。注意，空对象被序列化为空数组，反之亦然。

（4）GenericJackson2JsonRedisSerializer 类：使用 Jackson 实现的序列化器。该类实现 Java 对象到 JSON 字符串的序列化，以及 JSON 字符串到 Java 对象的反序列化。使用该类执行序列化时，会保存序列化的对象的包名和类名，反序列化时以包名和类名作为标识就可以还原成指定的对象。

（5）GenericToStringSerializer 类：该序列化器使用 Spring 的 ConversionService，使用默认的字符集 UTF-8 将对象转换为字符串，从而完成序列化。反之亦然。

（6）OxmSerializer 类：可实现对象与 XML 之间的相互转换。使用此序列化器，编程将会有些难度，而且效率最低，不建议使用。此外，该序列化器需要 Spring-OXM 模块的支持。

3. RedisTemplate 序列化相关的方法

RedisTemplate 类提供的与序列化相关的方法如表 3-8 所示。

表 3-8　RedisTemplate 类提供的与序列化相关的方法

方法原型	说　明
RedisSerializer <?> getDefaultSerializer()	获取当前模板（RedisTemplate）默认的序列化器
setDefaultSerializer(RedisSerializer <?> serializer)	设置用于当前模板的默认序列化器
RedisSerializer <?> getHashKeySerializer()	返回哈希键序列化器
void setHashKeySerializer(RedisSerializer <?> hashKeySerializer)	设置哈希键序列化器

续表

方法原型	说　明
RedisSerializer <?> getHashValueSerializer()	返回哈希值序列化器
void setHashValueSerializer(RedisSerializer <?> hashValueSerializer)	设置哈希值序列化器
RedisSerializer <?> getKeySerializer()	返回当前模板使用的键序列化器
void setKeySerializer(RedisSerializer <?> serializer)	设置当前模板使用的键序列化器
RedisSerializer < String > getStringSerializer()	返回字符串序列化器
void setStringSerializer(RedisSerializer < String > stringSerializer)	设置字符串序列化器
RedisSerializer <?> getValueSerializer()	返回当前模板使用的值序列化器
void setValueSerializer(RedisSerializer <?> serializer)	设置当前模板使用的值序列化器

4. RedisSerializer 接口的实现类的应用

下面结合例子介绍 serializer 包提供的几个 RedisSerializer 接口的实现类的应用。

（1）JdkSerializationRedisSerializer。

JdkSerializationRedisSerializer 序列化器是 RedisCache 类和 RedisTemplate 类默认的序列化器，采用 Java 语言的序列化机制。该序列化器将对象保存成二进制格式，执行序列化的效率不是最差的，但结果的可读性较差。下面的例子演示如何利用 JdkSerializationRedisSerializer 序列化器将一个对象进行序列化。

第一步，创建一个类 Barrel，代码如文件 3-52 所示。

【文件 3-52】 Barrel. java

```
1    @Data
2    public class Barrel implements Serializable {
3      private String material;
4      private double capacity;
5      //此处省略了 toString()方法
6    }
```

第二步，配置 RedisTemplate，代码如文件 3-53 所示。

【文件 3-53】 RedisSerializerConfig. java

```
1    @Configuration
2    public class RedisSerializerConfig {
3      //RedisConnectionFactory 配置元数据定义省略，代码见文件 3 - 6
4      @Bean
5      public RedisTemplate < String, Object > redisTemplate(
6          RedisConnectionFactory factory) {
7        RedisTemplate < String, Object > redisTemplate =
8          new RedisTemplate <>();
9        redisTemplate. setConnectionFactory(factory);
10       //指定值的默认的序列化器 JdkSerializationRedisSerializer
11       redisTemplate. setValueSerializer(RedisSerializer. java());
12       return redisTemplate;
13     }
14   }
```

第三步，编写测试代码，如文件 3-54 所示。

【文件 3-54】 TestSerializer.java

```
1    @RunWith(SpringJUnit4ClassRunner.class)
2    @ContextConfiguration(classes = RedisSerializerConfig.class)
3    public class TestSerializer {
4        @Autowired
5        private RedisTemplate<String, Object> template;
6        @Test
7        public void testJdkSerializer(){
8            Barrel barrel = new Barrel();
9            barrel.setMaterial("plastic");
10           barrel.setCapacity(60);
11           template.opsForValue().set("jdk",barrel);
12           System.out.println(template.opsForValue().get("jdk"));
13       }
14   }
```

运行此测试代码后，将 Barrel 对象序列化后保存到 Redis。控制台的输出为 Barrel{material='plastic', capacity=60.0}。利用 RedisInsight 客户端查看 Redis 保存的内容，如图 3-41 所示。

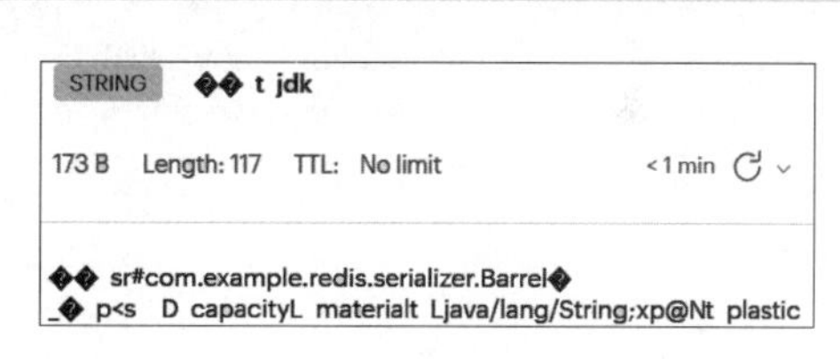

图 3-41 Redis 中保存的内容 1

(2) Jackson2JsonRedisSerializer。

为了解决图 3-41 中出现的乱码问题，可以替换默认的序列化接口实现机制。例如，使用 Jackson2JsonRedisSerializer 序列化器。Jackson2JsonRedisSerializer 是可以使用 Jackson 读取和写入 JSON 的序列化器。当要存储的值为字符串类型时，也可以采用 StringRedisSerializer 序列化器。本例采用 Jackson2JsonRedisSerializer 对 Barrel 对象进行序列化。

第一步，修改 RedisTemplate 类的默认序列化器，可利用表 3-8 中的 setKeySerializer()方法和 setValueSerializer()方法分别指定键和值的序列化器，以替代默认的序列化器。部分代码如文件 3-55 所示。

【文件 3-55】 RedisSerializerConfig.java

```
1    @Bean
2    public RedisTemplate<String,Object> redisTemplate(
3         RedisConnectionFactory factory) {
4       RedisTemplate<String,Object> template = new RedisTemplate<>();
5       template.setConnectionFactory(factory);
6       template.setKeySerializer(RedisSerializer.string());
7       template.setValueSerializer(new
8          Jackson2JsonRedisSerializer<Object>(Object.class));
9       template.afterPropertiesSet();
10      return template;
11   }
```

如文件 3-55 所示，第 6 行指定键的序列化器为 StringRedisSerializer，字符编码为默认的 UTF-8，第 7、8 行指定值的序列化器为 Jackson2JsonRedisSerializer。

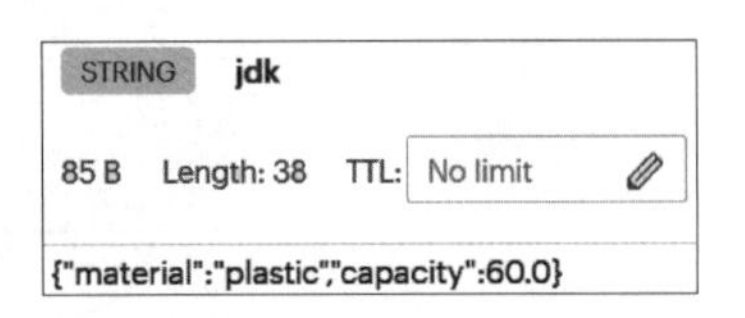

图 3-42　Redis 中保存的内容 2

第二步，引入 jackson-databind 依赖。当前的 Jackson2JsonRedisSerializer 并不要求持久化类(本例中为 Barrel 类)显式实现 Serializable 接口。运行文件 3-54 的测试代码后，控制台的输出为{material = plastic，capacity=60.0}，利用 RedisInsight 客户端查看 Redis 保存的内容，如图 3-42 所示。

(3) GenericJackson2JsonRedisSerializer。

该序列化器可以将 Java 对象序列化为 JSON 字符串，以及 JSON 字符串反序列化为 Java 对象。要使用 GenericJackson2JsonRedisSerializer 序列化器，可对文件 3-55 稍作修改，部分代码如文件 3-56 所示。

【文件 3-56】 RedisSerializerConfig. java 部分修改

```
1    @Bean
2    public RedisTemplate<String,Object> redisTemplate(
3      RedisConnectionFactory factory) {
4        RedisTemplate<String,Object> redisTemplate =
5           new RedisTemplate<>();
6        redisTemplate.setConnectionFactory(factory);
7        redisTemplate.setKeySerializer(RedisSerializer.string());
8        redisTemplate.setValueSerializer(RedisSerializer.json());
9        redisTemplate.afterPropertiesSet();
10       return redisTemplate;
11   }
```

GenericJackson2JsonRedisSerializer 序列化器并不要求持久化类显式实现 Serializable 接口。运行文件 3-54 中的测试代码后，控制台输出为 Barrel{material='plastic',capacity=60.0}，利用 RedisInsight 客户端查看 Redis 保存的内容，如图 3-43 所示。对比控制台输出和 Redis 中保存的内容可知，与 StringRedisSerializer 和 Jackson2JsonRedisSerializer 序列化器不同，在从 Redis 中获取键对应的值时，GenericJackson2JsonRedisSerializer 序列化器执行了反序列化操作。由 JSON 字符串反序列化为 Java 对象时，要求持久化类(本例中为 Barrel 类)提供公有的无参数构造方法。

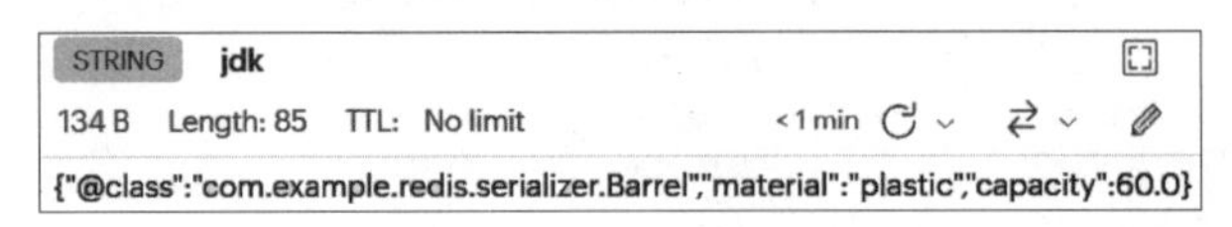

图 3-43　Redis 中保存的内容 3

3.12.2 HashMapper 接口

HashMapper 接口是 Spring Data Redis 提供的实现 Java 对象和 Redis Hash(Redis 的哈希类型)之间转换的核心接口。Redis Hash 一般具有如下结构：

```
Key:{
     filed: value,
     filed: value,
     filed: value,
     ....
}
```

这个结构和 Java 中的对象非常相似，但是不能按照 Java 对象的结构直接存储进 Redis Hash。因为 Java 对象中的字段(field)是可以嵌套的，而 Redis Hash 不支持嵌套结构。为此，Spring Data Redis 提供了将 Java 对象映射到 Redis Hash 的三种策略。

(1) 直接映射。可以使用 HashOperations 接口和相关的序列化程序，如 Jackson2JsonRedisSerializer，进行直接映射。

(2) 使用 Redis 仓库。Redis 仓库可以应用定制的映射策略转换和存储 Redis Hash 中的域对象，并且可以使用辅助索引(见 8.5 节)。

(3) 使用 HashMapper 接口和 HashOperations 接口。

本节主要介绍上述方法(3)执行 Java 对象与 Redis Hash 的映射。Spring Data Redis 提供了 HashMapper 接口的三个实现类，分别为 BeanUtilsHashMapper、ObjectHashMapper 和 Jackson2HashMapper。下面结合例子分别介绍 ObjectHashMapper 类和 Jackson2HashMapper 类的使用方法。

1. ObjectHashMapper 类

电商网站中经常有这样的需求：记录用户注册时填写的基础信息及其注册地址信息，实现这一需求的具体步骤如下。

第一步，创建两个持久化类 Person 和 Address，分别封装用户的基础信息和注册地址信息，代码如文件 3-57 和文件 3-58 所示。

【文件 3-57】 Person.java

```
1   @Data
2   public class Person {
3       private String personId;
4       private String firstname;
5       private String lastname;
6       private Address address;
7       private LocalDateTime localDateTime;
8   }
```

【文件 3-58】 Address.java

```
1   @Data
2   public class Address {
3     . private String city;
4       private String country;
5       //此处省略了带参数的构造方法
6   }
```

第二步，编写配置类，代码如文件 3-59 所示。

【文件 3-59】 HashMapperConfig.java

```
1   public class HashMapperConfig {
2       //此处省略了连接工厂的配置,内容见文件 3 - 6
3       @Bean
4       public RedisTemplate < String,Map < byte[ ],byte[ ]>>
5           hashMapperRedisTemplate(
```

```
6          RedisConnectionFactory factory) {
7          RedisTemplate < String, Map < byte[ ],byte[ ]>> template =
8             new RedisTemplate <>();
9          template.setConnectionFactory(factory);
10         //设置键序列化方式
11         template.setKeySerializer(RedisSerializer.string());
12         //设置简单类型值的序列化方式
13         template.setValueSerializer(RedisSerializer.byteArray());
14         //设置哈希类型键的序列化方式
15         template.setHashKeySerializer(RedisSerializer.byteArray());
16         //设置哈希类型值的序列化方式
17         template.setHashValueSerializer(RedisSerializer.byteArray());
18         template.afterPropertiesSet();
19         return template;
20     }
21  }
```

如文件 3-59 所示，第 11 行将键的序列化器指定为 StringRedisSerializer，默认的字符编码为 UTF-8。第 12～17 行分别将字符串类型值、哈希键和哈希值的序列化器指定为 ByteArrayRedisSerializer，该序列化器是一个只使用字节数组（byte[]）的原始序列化器（RedisSerializer），代码如下：

```
1   package org.springframework.data.redis.serializer;
2   enum ByteArrayRedisSerializer implements RedisSerializer < byte[ ]> {
3       INSTANCE;
4   
5       @Nullable
6       @Override
7       public byte[ ] serialize(@Nullable byte[ ] bytes)
8          throws SerializationException {
9            return bytes;
10      }
11  
12      @Nullable
13      @Override
14      public byte[ ] deserialize(@Nullable byte[ ] bytes)
15         throws SerializationException {
16           return bytes;
17      }
18  }
```

第三步，建立 person 对象与 Redis Hash 的映射，编写的测试代码如文件 3-60 所示。

【文件 3-60】 TestHashMapper.java

```
1   @RunWith(SpringJUnit4ClassRunner.class)
2   @ContextConfiguration(classes = HashMapperConfig.class)
3   public class TestHashMapper {
4      @Autowired
5      private RedisTemplate < String, Map < byte[ ],byte[ ]>> redisTemplate;
6   
```

```
7        private final HashMapper<Object, byte[], byte[]> hashMapper =
8          new ObjectHashMapper();
9
10       @Test
11       public void save(){
12          Person person = new Person();
13          person.setPersonId("person-address");
14          person.setFirstname("Simth");
15          person.setLastname("Qiong");
16          person.setLocalDateTime(LocalDateTime.now());
17          person.setAddress(new Address("DaLian","China"));
18          Map<byte[], byte[]> map = hashMapper.toHash(person);
19          HashOperations<String,byte[],byte[]> ops =
20            redisTemplate.opsForHash();
21          ops.putAll(person.getPersonId(),map);
22          map.entrySet().iterator().forEachRemaining(entry ->
23             System.out.println(new String(entry.getKey()) + " = " +
24             new String(entry.getValue())));
25       }
26    }
```

如文件3-60所示，第11行开始的测试用例是实现将person对象映射为Redis Hash并存入Redis。因此，需要调用HashMapper接口的toHash()方法。该方法的原型为：

```
Map<K,V> toHash(T object)
```

其功能是将Java对象（本例中为person对象）转换为可以被Redis Hash使用的映射（第18行）。同时，指定了Redish Hash的字段和值的类型均为byte[]（字节数组）。与此对应的是第7、8行，在创建ObjectHashMapper的实例时，指明HashMapper的三个泛型，第一个为Java对象的类型，第二个和第三个为Redis Hash中的字段和值的类型。第19～21行将person对象映射成的map对象存入Redis，并且约定，该map对象的键的类型为String。作为测试，第22～24行将转换后的map对象的内容在控制台输出。运行此测试代码后，控制台输出和Redis中保存的内容分别如图3-44和图3-45所示。

```
_class=com.example.redis.hashmapper.entity.Person
address.city=DaLian
address.country=China
firstname=Simth
lastname=Qiong
localDateTime=2023-02-03T16:15:26.742473400
personId=person-address
```

图3-44　控制台输出的内容

Field	Value
_class	com.example.redis.hashmapper...
address.city	DaLian
address.country	China
firstname	Simth
lastname	Qiong
localDateTime	2023-02-03T16:15:26.742473400
personId	person-address

图3-45　Redis中保存的内容4

第四步，验证Redis Hash向Java对象的（反向）映射。可调用HashMapper接口的fromHash()方法实现（反向）映射。该方法的原型为：

```
T fromHash(Map<K,V> hash)
```

可在文件3-60的基础上增加一个测试用例，代码如下：

```
1   @Test
2   public void find() {
3     HashOperations<String, byte[], byte[]> ops =
4       redisTemplate.opsForHash();
5     Map<byte[], byte[]> map = ops.entries("person-address");
6     Person person = (Person)hashMapper.fromHash(map);
7     System.out.println(new Gson().toJson(person));
8   }
```

运行此测试用例，可在控制台看到以 JSON 字符串形式输出的 person 对象，如图 3-46 所示。

```
{"personId":"person-address","firstname":"Simth","lastname":"Qiong",
"address":{"city":"DaLian","country":"China"},
"localDateTime":{"date":{"year":2023,"month":2,"day":3},
"time":{"hour":16,"minute":15,"second":26,"nano":742473400}}}
```

图 3-46 控制台输出的 JSON 字符串

2. Jackson2HashMapper 类

Jackson2HashMapper 类通过使用 Faster XML Jackson 为 Java 对象提供 Redis Hash 映射。Jackson2HashMapper 可以将顶级属性映射为哈希字段名，还可以选择将结构扁平化。扁平化是指为所有嵌套属性创建单独的哈希项（字段和值），并尽可能将复杂类型解析为简单类型。以文件 3-57 定义的持久化类为例，数据在非扁平化映射中的显示方式如表 3-9 所示。

表 3-9 数据在非扁平化映射中的显示方式

字 段	值
firstname	Jon
lastname	Snow
address	{ "city" : "Castle Black", "country" : "The North" }
localDateTime	2018-01-02T12:13:14

经过扁平化处理后，显示方式如表 3-10 所示。

表 3-10 数据在扁平化映射中的显示方式

字 段	值
firstname	Jon
lastname	Snow
address.city	Castle Black
address.country	The North
localDateTime	2018-01-02T12:13:14

对于文件 3-57 定义的 Person 类的对象，也可以用 Jackson2JsonRedisSerializer 类完成其与 Redis Hash 的映射。

第一步，可以修改文件 3-59，配置 RedisTemplate 的相应的序列化器。部分代码如文件 3-61 所示。

【文件 3-61】 HashMapperConfig. java 的部分修改

```
1   @Bean
2   public RedisTemplate<String, Map<String,Object>>
3     hashMapperRedisTemplate(RedisConnectionFactory factory) {
4       RedisTemplate<String,Map<String,Object>> redisTemplate =
5           new RedisTemplate<>();
6       redisTemplate.setConnectionFactory(factory);
7       // 设置键序列化方式
8       redisTemplate.setKeySerializer(RedisSerializer.string());
9       redisTemplate.setValueSerializer(new
10        Jackson2JsonRedisSerializer<Object>(Object.class));
11      redisTemplate.setHashKeySerializer(new
12        Jackson2JsonRedisSerializer<Object>(Object.class));
13      redisTemplate.setHashValueSerializer(new
14        Jackson2JsonRedisSerializer<Object>(Object.class));
15      redisTemplate.afterPropertiesSet();
16      return redisTemplate;
17  }
```

第二步，执行 person 对象到 Redis Hash 的映射，部分测试代码如文件 3-62 所示。

【文件 3-62】 部分测试代码

```
1   @Test
2   public void save(){
3     Person person = new Person();
4     person.setPersonId("person-address");
5     person.setFirstname("Simth");
6     person.setLastname("Qiong");
7     person.setLocalDateTime(LocalDateTime.now());
8     person.setAddress(new Address("DaLian","China"));
9     HashMapper<Object, String, Object> hashMapper =
10      new Jackson2HashMapper(true);
11    Map<String, Object> map = hashMapper.toHash(person);
12    HashOperations<String,String,Object> ops =
13      redisTemplate.opsForHash();
14    ops.putAll(person.getPersonId(),map);
15    map.entrySet().iterator().forEachRemaining(entry ->
16      System.out.println(entry.getKey() + " = " + entry.getValue()));
17  }
```

如文件 3-62 所示，第 9、10 行实例化 Jackson2HashMapper 映射器，该映射器的构造方法有布尔型参数，取值为 true 意味着采用扁平化方式处理数据；反之取值为 false。此外，Jackson2HashMapper 类实现的 HashMapper<K,T,V>接口中，已将 K、T 和 V 三个泛型指定为 Object、String、Object。即该映射器的定义为：

```
1   public class Jackson2HashMapper implements HashMapper<Object, String,
2     Object>, HashObjectReader<String, Object> {
3         ……
4   }
```

因此，实例化该映射器时必须沿用该映射器的泛型定义。为了便于处理 LocalDateTime 类型的数据，需要在 pom. xml 文件中增加依赖：

```
1   <dependency>
2       <groupId>com.fasterxml.jackson.datatype</groupId>
3       <artifactId>jackson-datatype-jsr310</artifactId>
4       <version>2.13.3</version>
5   </dependency>
```

第三步，执行此测试代码，控制台的输出如图 3-47 所示。

```
localDateTime=2023-02-03T20:27:58.7118286
firstname=Simth
@class=com.example.redis.hashmapper.entity.Person
address.city=DaLian
personId=person-address
lastname=Qiong
address.@class=com.example.redis.hashmapper.entity.Address
address.country=China
```

图 3-47　控制台的输出

第四步，执行 Redis Hash 到 person 对象的映射。部分测试代码如文件 3-63 所示。

【文件 3-63】部分测试代码

```
1   @Test
2   public void find() {
3     HashOperations<String,String,Object> ops =
4       redisTemplate.opsForHash();
5     Map<String,Object> map = ops.entries("person-address");
6     HashMapper<Object, String, Object> hashMapper =
7       new Jackson2HashMapper(true);
8     Person person = (Person)hashMapper.fromHash(map);
9     System.out.println(new Gson().toJson(person));
10  }
```

注意，为保证反序列化能够成功执行，需要 Person 类和 Address 类都提供公有的无参数构造方法。

3.13　小结

本章着重介绍了 Spring Data Redis 提供的 RedisTemplate 类。RedisTemplate 类是在 RedisConnection 接口（org. springframework. data. redis. connection. RedisConnection）基础上，将 Redis 操作进行了更高层次的封装。由于 RedisTemplate 来自于 Spring，因此可以在程序中利用 IoC、AOP 等 Spring 的特性优雅地、简单地操作 Redis。本章涉及的操作 Redis 的常用类和接口的关系如图 3-48 所示。

对于操作 Redis，可以使用 Lettuce、Jedis、Redisson 等客户端，也可以利用 Spring 框架的 RedisConnectionFactory 接口创建与 Redis 的连接，再使用 RedisTemplate 类操作 Redis。图 3-48 只列举了 RedisTemplate 类的一些方法。如果在 Redis 中保存的键和值都是字符串类型，也可以使用 RedisTemplate 类的子类 StringRedisTemplate。

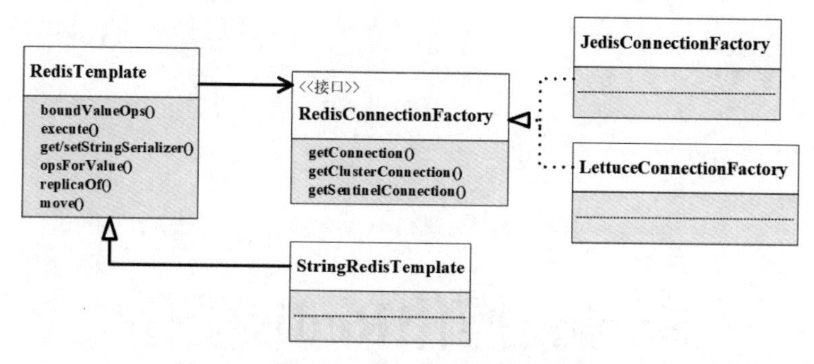

图 3-48 操作 Redis 的常用类和接口的关系

作为 Spring Data Redis 提供的模板类，RedisTemplate 提供了大量的 API 用以封装 Redis 操作。如，数据的序列化和反序列化操作、执行 Lua 脚本(见 5.4.2 节)、操作 Redis 分片集群(见 7.3.3 节)等。同时 RedisTemplate 类还提供了操作字符串、列表、哈希等数据结构的专属操作接口。可以说，RedisTemplate 类是使用 Spring 开发 Redis 应用的首选。

第 4 章

Spring操作Redis缓存

在一些高并发的应用场景里,如果读写数据的请求都集中到 MySQL 等关系数据库上,会造成关系数据库主机性能下降甚至宕机,从而导致严重的问题。此时就有必要引入基于内存读写的 Redis 来缓存数据以减轻 MySQL 等关系数据库的读写压力。

本章将模拟项目需求,给出 Spring 整合 MySQL 和 Redis 的开发范例。

4.1 JdbcTemplate

4.1.1 JdbcTemplate 简介

JdbcTemplate(org.springframework.jdbc.core.JdbcTemplate)是 org.springframework.jdbc.core 包中的核心类,是 Spring 对 JDBC(Java DataBase Connectivity,Java 数据库连接)API 的封装。该类可以执行 SQL 查询和更新(添加、修改和删除)操作,对结果集(ResultSet)执行遍历,并且可以捕获 JDBC 异常并将其转换为 org.springframework.dao 包中定义的异常。

对于数据库操作,不管使用什么样的技术,都需要一些特定的步骤。例如,都需要获取一个数据库连接,并在操作完成后释放资源。这些都是数据库操作中的固定步骤,但是每次数据库操作方法又有些不同,即可能会查询不同的对象或以不同方式更新数据,这些都是数据库操作中变化的部分。

Spring 将数据库操作中固定的和变化的部分划分为两个类:模板(Template)和回调(Callback)。模板管理数据库操作过程中固定的部分,而回调处理自定义的数据访问代码。图 4-1 展示了这两个类的职责。

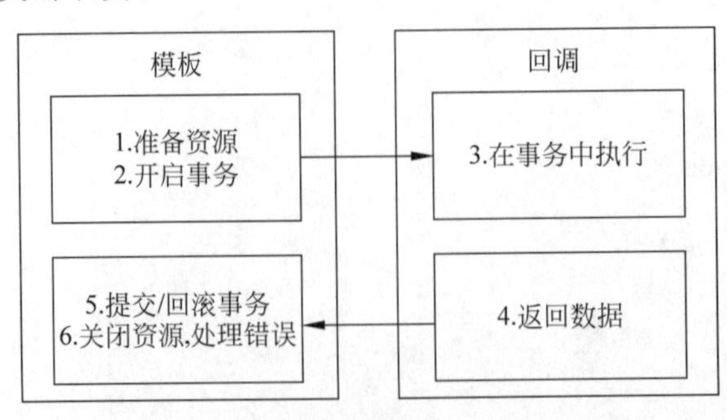

图 4-1 模板与回调

如图 4-1 所示，Spring 的模板类处理数据库操作的固定部分——事务控制、管理资源以及处理异常。同时，与应用程序相关的数据访问——SQL 语句、绑定参数以及整理结果集在回调方法中处理。这是一个优雅的设计方案，开发人员只需关注与应用程序相关的数据访问部分即可。针对不同的持久化平台，Spring 提供了多个可选的模板。如果使用 JPA（Java Persistence API）技术，可以选择 Spring Data JPA 模块的 JpaRepository 接口。如果使用 JDBC，Spring 提供了如下两个模板类供选择。

（1）JdbcTemplate 类：最基本的 JDBC 模板，这个模板支持简单的 JDBC 数据库访问功能以及基于索引参数的查询。

（2）NamedParameterJdbcTemplate 类：使用该模板类执行查询时可以将值以命名参数的形式绑定到 SQL 中，而不是使用简单的索引参数。

对于开发者来讲，只有在需要命名参数时才使用 NamedParameterJdbcTemplate。因此，对于大多数 JDBC 开发任务来说，JdbcTemplate 就是最好的方案。

4.1.2 JdbcTemplate 的常用方法

JdbcTemplate 提供了大量的查询和更新数据库的方法，常用方法如表 4-1 所示。

表 4-1 JdbcTemplate 的常用方法

方 法	说 明
public <T> List<T> query(String sql,RowMapper<T> rowMapper,@Nullable Object... args)	其中，参数 sql 代表用于执行查询的语句；参数 rowMapper 可以将查询结果集中的每一行数据映射为一个对象；参数 args 表示需要传入 SQL 语句的参数
public <T> T queryForObject(String sql,RowMapper<T> rowMapper,@Nullable Object... args)	
public int update(String sql)	用于执行添加、修改、删除等语句。其中，参数 sql 代表需要执行的 SQL 语句；参数 args 代表需要传入 SQL 语句的参数
public int update(String sql,Object... args)	
public void execute(String sql)	可以执行任意 SQL 语句，一般用于执行 DDL 语句。其中，参数 sql 代表需要执行的 SQL 语句；参数 action 代表执行完 SQL 语句后要调用的函数
public T execute(String sql, PreparedStatementCallback action)	
public int[] batchUpdate(String sql,List<Object[]> batchArgs,final int[]argTypes)	用于批量执行添加、修改、删除等语句。其中，参数 sql 代表需要执行的 SQL 语句；参数 batchArgs 代表需要传入 SQL 语句的参数；参数 argTypes 代表 batchArgs 的 JDBC 类型

在社交网络中，经常有这样的需求：在用户的个人主页上显示该用户的发文数或作品总数。可通过 Spring JdbcTemplate 操作数据库实现这一需求，具体步骤如下。

1. 创建项目

创建一个名为 spcache 的 Maven 项目。并加入相关依赖。需要引入的 jar 包有 Spring-jdbc、mysql-java-connector、Spring-test 等。可在 2.3 节项目的基础上增加以下依赖：

```
1    <dependency>
2        <groupId>mysql</groupId>
3        <artifactId>mysql-connector-java</artifactId>
4        <version>8.0.28</version>
```

```
5           </dependency>
6           <!-- Spring Test -->
7           <dependency>
8               <groupId>org.springframework</groupId>
9               <artifactId>spring-test</artifactId>
10              <version>${spring.version}</version>
11          </dependency>
12          <!-- Spring JDBC Template -->
13          <dependency>
14              <groupId>org.springframework</groupId>
15              <artifactId>spring-jdbc</artifactId>
16              <version>${spring.version}</version>
17          </dependency>
18          <!-- Spring 日志包 slf4j&logback -->
19          <dependency>
20              <groupId>ch.qos.logback</groupId>
21              <artifactId>logback-classic</artifactId>
22              <version>1.2.10</version>
23          </dependency>
24          <!-- redis -->
25          <dependency>
26              <groupId>redis.clients</groupId>
27              <artifactId>jedis</artifactId>
28              <version>3.8.0</version>
29          </dependency>
30          <dependency>
31              <groupId>org.springframework.data</groupId>
32              <artifactId>spring-data-redis</artifactId>
33              <version>2.7.1</version>
34          </dependency>
```

2. 创建数据库表单

创建一个名为 spring 的数据库，并在 spring 数据库中创建数据表 tb_user，创建表单的语句如下：

```
CREATE TABLE `tb_user` (
  `user_id` int NOT NULL AUTO_INCREMENT COMMENT '用户 ID',
  `user_name` varchar(50) DEFAULT NULL COMMENT '用户名',
  `balance` bigint DEFAULT 0 COMMENT '作品数',
  PRIMARY KEY (`user_id`)
) ENGINE = InnoDB AUTO_INCREMENT = 1 DEFAULT CHARSET = utf8mb3;
```

3. 编写配置文件

在 src/main/resources 目录下创建一个名为 jdbc.properteis 的配置文件，设置数据库连接的关键参数，内容如文件 4-1 所示。

【文件 4-1】 jdbc.properties

```
1    jdbc.driver = com.mysql.cj.jdbc.Driver
2    jdbc.url = jdbc:mysql://127.0.0.1:3306/spring
3    jdbc.username = root
4    jdbc.password = password
```

在 src/main/java 目录下创建名为 spring.jdbc 的包，在该包中创建 Spring 的配置类，用于声明 Spring 中的 Bean，内容如文件 4-2 所示。

【文件 4-2】 SpringConfig.java

```
1    package spring.jdbc;
2    //import 部分略
3    @Configuration
4    @PropertySource("classpath:jdbc.properties")
5    @ComponentScan(basePackages = "spring.jdbc")
6    public class SpringConfig {
7        @Value("${jdbc.driver}")
8        private String driver;
9        @Value("${jdbc.url}")
10       private String url;
11       @Value("${jdbc.username}")
12       private String username;
13       @Value("${jdbc.password}")
14       private String password;
15
16       @Bean
17       public DriverManagerDataSource dataSource(){
18           DriverManagerDataSource ds = new DriverManagerDataSource();
19           ds.setDriverClassName(driver);
20           ds.setUrl(url);
21           ds.setUsername(username);
22           ds.setPassword(password);
23           return ds;
24       }
25
26       @Bean
27       public JdbcTemplate jdbcTemplate(@Autowired
28             DriverManagerDataSource ds){
29           JdbcTemplate template = new JdbcTemplate();
30           template.setDataSource(ds);
31           return template;
32       }
33   }
```

如文件 4-2 所示，第 4 行指定 classpath 中的配置文件 jdbc.properties。第 5 行指定了组件扫描的基础包为 spring.jdbc，这就意味着需要将 Spring 中的 Bean 存放在该包或该包的子包中，并用相应的注解标注（见文件 4-5 和文件 4-6）。第 7～14 行读取 jdbc.properties 文件的内容并完成相应的属性赋值。第 16～24 行配置数据源 Bean（DriverManagerDataSource Bean）；第 26～32 行配置 JdbcTemplate。

4. 创建实体类

在 src/main/java 文件夹下创建一个名为 spring.jdbc.entity 的包，并在该包中创建一个名为 User 的类，代码如文件 4-3 所示。

【文件 4-3】 User.java

```
1    package spring.jdbc.entity;
2
3    public class User {
```

```
4        private long userId;
5        private String userName;
6        private Long balance;
7        //此处省略了 getters/setters 方法
8        //此处省略了 toString()方法
9    }
```

5. 创建数据访问组件

在 src/main/java 目录下创建名为 spring. jdbc. dao 的包,并创建数据访问接口 UserDao 和对应的实现类 UserDaoImpl。其中,UserDao 代码如文件 4-4 所示。

【文件 4-4】 UserDao. java

```
1    package spring.jdbc.dao;
2
3    public interface UserDao {
4        public int add(User user);
5        public int update(User user);
6        public int count();
7        public List<User> getList();
8        public User getUser(Integer id);
9        public void batchAddUser(List<Object[]> users);
10   }
```

UserDaoImpl 使用 JDBC Template 方式实现数据库操作,该类被注解@Repository 标注,将其注册为 Spring 的 Bean,代码如文件 4-5 所示。

【文件 4-5】 UserDaoImpl. java

```
1    package spring.jdbc.dao.impl;
2    //import 部分略
3    @Repository("userDao")
4    public class UserDaoImpl implements UserDao{
5        @Autowired
6        private JdbcTemplate template;
7
8        public int add(User user){
9            String sql = "insert into tb_user(user_name,balance)
10               values(?,?);";
11           int update = template.update(
12               sql,user.getUserName(),user.getBalance());
13           return update;
14       }
15
16       public int update(User user){
17           String sql = "update tb_user set user_name = ?, balance = ?"
18           + " where user_id = ?";
19           return template.update(sql,user.getUserName(),
20               user.getBalance(),user.getUserId());
21       }
22
23       public int delete(String name){
```

```
24              String sql = "delete from tb_user where user_name = ?;";
25              return template.update(sql,name);
26          }
27
28          public int count(){
29              String sql = "select count( * ) from tb_user;";
30              return template.queryForObject(sql, Integer.class);
31          }
32
33          public List < User > getList(){
34            String sql = "SELECT user_id,user_name,balance from tb_user;";
35            return template.query(sql, new
36                 BeanPropertyRowMapper < User >(User.class));
37          }
38
39          public User getUserById(long id){
40              try{
41                  String sql = "SELECT user_id,user_name,balance" +
42                      "from tb_user where user_id = ?;";
43                  return template.queryForObject(sql,
44                      new BeanPropertyRowMapper < User >(User.class), id);
45              } catch(EmptyResultDataAccessException ex) {
46                  return null;
47              }
48          }
49
50          public void batchAddUser(List < Object[ ]> users){
51              String sql = "INSERT into tb_user(user_name, balance)
52                  VALUES(?,?);";
53             template.batchUpdate(sql, users);
54          }
55      }
```

在文件4-5中，第8～14行实现添加功能，第16～21行实现更改功能，第28～31行实现用户数量统计功能，第33～37行实现返回所有记录功能，第39～48行实现根据id查找用户信息功能，第50～54行实现批量添加功能。其中，BeanPropertyRowMapper类(第36、44行)是RowMapper接口的实现类，它可以自动将数据表中的数据映射到用户自定义的类中(前提是用户自定义类中的属性要与数据表中的字段项对应)。

6. 编写服务组件

在src/main/java目录下创建一个名为spring.jdbc.service的包，并创建服务组件接口UserService，该接口定义的抽象方法与UserDao接口中定义的抽象方法相同(见文件4-4)，此处略。创建UserService接口的实现类UserServiceImpl，代码如文件4-6所示。

【文件4-6】 UserServiceImpl.java

```
1   package spring.jdbc.service.impl;
2   //import 部分略
3   @Service("userService")
4   public class UserServiceImpl implements UserService {
5
```

```
6       @Autowired
7       private UserDao userDao;
8
9       public int add(User user){
10          return userDao.add(user);
11      }
12      public User getUserById(long id){
13          return userDao.getUserById(id);
14      }
15      //此处省略了对 UserService 接口的其余方法的调用
16  }
```

7. 编写测试类

在 src/test/java 目录下创建测试类,以下给出测试批量添加功能的测试代码,如文件 4-7 所示。

【文件 4-7】 TestJdbcTemplate.java

```
1   package spring.jdbc.test;
2   //import 部分略
3   @RunWith(SpringJUnit4ClassRunner.class)
4   @ContextConfiguration(classes = {SpringConfig.class,
5      RedisConfig.class})
6   public class TestJdbcTemplate {
7      @Autowired
8      private UserService us;
9
10     @Test
11     public void testAddBatchUser(){
12         List<Object[]> users = new ArrayList<Object[]>();
13         Object[] u11 = {"Jetty", 201L};
14         Object[] u12 = {"Peter", 1234L};
15         Object[] u13 = {"Linda", 543L};
16         Object[] u14 = {"Momot", 450L};
17         users.add(u11);
18         users.add(u12);
19         users.add(u13);
20         users.add(u14);
21         us.batchAddUser(users);
22     }
23  }
```

运行批量添加功能的测试代码。在 MySQL 客户端查看运行结果,如图 4-2 所示。

```
mysql> select * from tb_user;
+---------+-----------+---------+
| user_id | user_name | balance |
+---------+-----------+---------+
|       1 | Jetty     |     201 |
|       2 | Peter     |    1234 |
|       3 | Linda     |     543 |
|       4 | Momot     |     450 |
+---------+-----------+---------+
4 rows in set (0.00 sec)
```

图 4-2 批量添加运行结果

4.2 Spring 整合 Redis 缓存

Spring 自身并没有缓存，但它对缓存提供了声明式的支持，能够与多种流行的缓存实现集成，如 Ehcache、Redis、JCache 等。Spring 对缓存的支持有两种方式：注解驱动式缓存和 XML 声明式缓存。从实现形式来讲，缓存是一个键值对，其中键一般为字符串类型，值可以是各种数据类型。Redis 作为键值型数据库，非常适合做缓存。本节关注如何应用 Spring 集成 Redis 缓存。

为实现声明式缓存，Spring 提供了一个 CacheManager 接口（缓存管理器，org.springframework.cache.CacheManager）。缓存管理器是 Spring 抽象缓存的核心，它能够与多个流行的缓存实现集成。对于 Redis，Spring Data Redis 提供了 RedisCacheManager 类（org.springframework.data.redis.cache.RedisCacheManager），它是 CacheManager 接口的实现类。RedisCacheManager 会与一个 Redis 服务器协作，并通过 RedisTemplate 类将缓存条目存储到 Redis 中。通过 RedisCacheManager 集成 Redis 缓存可遵循以下步骤。

1. 配置 Redis 缓存管理器

为了使用 RedisCacheManager，需要一个 RedisTemplate Bean 以及一个 RedisConnectionFactory 实现类（如 JedisConnectionFactory）的 Bean。此外，要启动 Spring 对注解驱动的缓存管理功能，需要在配置类上添加一个@EnableCaching 注解。随后，配置 RedisCacheManager 就非常简单了，如文件 4-8 所示。

【文件 4-8】 RedisConfig.java

```
1    package spring.jdbc;
2    //import 部分略
3    @Configuration
4    @EnableCaching
5    public class RedisConfig {
6      @Bean
7      public RedisConnectionFactory redisConnectionFactory() {
8        JedisPoolConfig jpc = new JedisPoolConfig();
9        jpc.setMaxTotal(8);
10       jpc.setMaxIdle(8);
11       jpc.setMinIdle(0);
12       jpc.setMaxWait(Duration.ofMillis(200));
13
14       //Redis 连接配置
15       RedisStandaloneConfiguration redisStandaloneConfiguration =
16           new RedisStandaloneConfiguration();
17       //设置连接的 ip
18       redisStandaloneConfiguration.setHostName("127.0.0.1");
19       //端口号
20       redisStandaloneConfiguration.setPort(6379);
21       //连接的数据库
22       redisStandaloneConfiguration.setDatabase(0);
23       //JedisConnectionFactory 配置 jedisPoolConfig
24       JedisClientConfiguration.JedisClientConfigurationBuilder
25           jedisClientConfiguration = JedisClientConfiguration.builder();
26       //连接池
```

```
27          jedisClientConfiguration.usePooling().poolConfig(jpc);
28          //工厂对象
29          RedisConnectionFactory factory = new JedisConnectionFactory(
30            redisStandaloneConfiguration, jedisClientConfiguration.build());
31                return factory;
32          }
33
34          @Bean
35          public StringRedisTemplate redisTemplate(@Autowired
36            RedisConnectionFactory rcf){
37              StringRedisTemplate template = new StringRedisTemplate();
38              template.setConnectionFactory(rcf);
39              return template;
40          }
41
42          @Bean
43          public RedisTemplate<String, User> template(@Autowired
44            RedisConnectionFactory rcf){
45              RedisTemplate<String, User> t =
46                  new RedisTemplate<String, User>();
47              t.setConnectionFactory(rcf);
48              return t;
49          }
50
51          @Bean(name = {"redisCache1","redisCache2"})
52          public CacheManager cacheManager(@Autowired
53            RedisConnectionFactory rcf) {
54              return RedisCacheManager.create(rcf);
55          }
56      }
```

为了将从数据库中读取的 User 对象缓存到 Redis，文件 4-8 创建了一个普通的 RedisTemplate 对象（第 42～49 行），同时指定了 RedisTemplate 的键和值的类型分别为 String 和 User。第 51～55 行通过给 RedisCacheManager 类的构造方法传递一个 RedisConnectionFactory 对象作为参数，创建了两个 RedisCacheManager 对象：redisCache1 和 redisCache2。

2. 修改持久化类

向 Redis 中存储对象要进行序列化，需要将对象转换为一种可跨平台识别的字节格式。目标平台可以通过字节信息解析还原对象，即反序列化。在本例中，要在 Redis 中缓存 User 对象数据，需要 User 类显式实现 java.io.Serializable 接口。

3. 实现注解声明式缓存

Spring 中应用于缓存的注解共有 5 个，如表 4-2 所示。

表 4-2 Spring 应用于缓存的注解

注解	描 述
@Cacheable	在调用方法之前，Spring 先到缓存服务器中查找对应键的缓存值，如果找到缓存值，那么 Spring 将不会再调用该注解标注的方法，而是将缓存值读出，返回给调用者；如果没有找到缓存值，那么 Spring 就会执行该注解标注的方法，并将该方法的返回结果通过键保存到缓存服务器中

续表

注解	描　述
@CachePut	Spring 会将该注解标注的方法的返回值写入缓存服务器中。需要注意的是，Spring 不会事先去缓存服务器中查找，而是直接执行方法，然后缓存返回值。即该注解标注的方法始终会被 Spring 调用
@CacheEvict	移除缓存对应的键值
@Caching	分组注解，它能够同时应用于其他缓存的注解
@CacheConfig	在类级别共享一些常见的缓存设置

一般而言，对于查询操作，可使用@Cacheable 注解；对于添加和修改操作，可使用@CachePut 注解；对于删除操作，可使用@CacheEvict 注解。@Cacheable 注解和@CachePut 注解都可以保存键值对，只是方式略有不同。它们只能运用于有返回值的方法。而删除缓存键的@CacheEvict 注解则可以用在返回值为 void 的方法上，因为它并不需要保存任何值。

上述注解都能标注到类或者方法上。如果标注到类上，则对类的所有方法都有效；如果标注到方法上，则只对被标注的方法有效。大部分情况下，会标注到方法上。例如，可将表 4-2 中的注解应用于文件 4-6 的部分方法上，代码如文件 4-9 所示。

【文件 4-9】 UserServiceImpl.java

```
1   package spring.jdbc.service.impl;
2   //import 部分略
3   @Service("userService")
4   public class UserServiceImpl implements UserService {
5       @Autowired
6       private UserDao userDao;
7
8       @CachePut(value = "redisCache1", key = "15")
9       public int update(User user){
10          return userDao.update(user);
11      }
12
13      @Caching(evict = { @CacheEvict(value = "redisCache1",
14          key = "'redis_user_1'"), @CacheEvict(
15          cacheNames = "redisCache2",allEntries = true) })
16      public int delete(String name){
17          return userDao.delete(name);
18      }
19
20      @Cacheable(value = "redisCache1")
21      public List<User> getList(){
22          System.out.println("========= DataBase =========");
23          return userDao.getList();
24      }
25
26      @Cacheable(cacheNames = {"redisCache1", "redisCache2"},
27          key = "'redis_user_' + #id", condition = "#id < 3 ")
28      public User getUserById(long id){
29          return userDao.getUserById(id);
30      }
31  }
```

其中，第20行应用@Cacheable注解将getList()方法的返回值存入缓存redisCache1。由value属性指定的缓存管理器为文件4-8中创建的缓存管理器（见文件4-8第51～55行）。第22行增加控制台输出，用于检验该方法是否被调用。类似地，第26、27行应用@Cacheable注解将getUserById()方法的返回值存入缓存redisCache1和redisCache2，其中cacheNames属性和value属性表达同样的意思，但这两个属性不能同时使用。由key属性指定缓存数据的键为redis_user_#id，其中#id是Spring EL表达式，表示id的值由getUserById()方法的参数id指定；condition属性则指示执行缓存的条件，本例中只缓存id值小于3的数据。第8行应用@CachePut注解将update()方法的运行结果（本例为更改信息后的User对象）写入缓存redisCache1中，同时方便查看结果，将其键设置为一个常量。第13～15行应用@Caching注解标注delete()方法，实现在数据库中删除某条记录时，删除两个缓存中相应的内容。

4. 运行测试

（1）读取数据表tb_user中的所有记录，并将它们存入redisCache1缓存。实现代码为文件4-9的第20～24行。编写对应的测试代码。

```
1    @Test
2    public void testGetList() {
3        us.getList().forEach(System.out::println);
4    }
```

运行测试代码后的输出如图4-3所示。

因为getList()方法被@Cacheable注解标注，所以该方法的返回值会被存入redisCache1缓存中。当再次执行相同条件的查询请求时，由于查询结果数据集已保存在Redis缓存中，则无须再次执行数据库查询，而是直接从缓存中获取数据，即getList()方法不会被调用。再次执行查询时的控制台输出如图4-4所示。

```
C:\software\jdk-11.0.13\bin\java.exe ...
========= DataBase =========
User [userId=1, username=Jetty, balance=201.43]
User [userId=2, username=Peter, balance=1234.32]
User [userId=3, username=Linda, balance=543.75]
User [userId=4, username=Momot, balance=450.31]
```

图4-3　首次执行查询时的控制台输出

```
C:\software\jdk-11.0.13\bin\java.exe ...
User [userId=1, username=Jetty, balance=201.43]
User [userId=2, username=Peter, balance=1234.32]
User [userId=3, username=Linda, balance=543.75]
User [userId=4, username=Momot, balance=450.31]
```

图4-4　再次执行查询时的控制台输出

（2）从数据表tb_user中取出id值为1的记录，并将数据存入redisCache1缓存和redisCache2缓存，键为redis_user_1。实现代码为文件4-9的第26～30行。编写测试代码并运行，利用Redis客户端查看运行结果如图4-5所示。

（3）将id为3的用户的名字改为Lin_Da，并查看redisCache1缓存的变化情况。实现代码为文件4-9的第8～11行。编写测试代码并运行，利用Redis客户端查看运行结果如图4-6所示。

```
127.0.0.1:6379> keys *
1) "redisCache1::redis_user_1"
2) "redisCache2::redis_user_1"
3) "redisCache1::SimpleKey []"
```

图4-5　利用Redis客户端查看(2)的运行结果

```
127.0.0.1:6379> keys *
1) "redisCache1::redis_user_1"
2) "redisCache1::15"
3) "redisCache2::redis_user_1"
4) "redisCache1::SimpleKey []"
```

图4-6　利用Redis客户端查看(3)的运行结果

(4) 从数据表 tb_user 中删除 id 为 1 的用户数据，并同步移除缓存中的数据。在执行测试(2)时已将该用户的信息分别存入 redisCache1 和 redisCache2 缓存。为此，在移除缓存数据时，对两个缓存采取了不同的策略。其中，对于 redisCache1 缓存，移除键为'redis_user_1'的条目；对于 redisCache2 缓存，移除该缓存中的所有条目，实现代码为文件 4-9 的第 13～18 行。编写测试代码并执行，利用 Redis 客户端查看运行结果如图 4-7 所示。类似地，可以将文件 4-9 的第 13～15 行的注解替换为@CacheEvict(cacheNames = "redisCache1",allEntries = true)，用于在删除数据时清空 redisCache1 缓存。

```
127.0.0.1:6379> keys *
1) "redisCache1::15"
2) "redisCache1::SimpleKey []"
```

图 4-7 利用 Redis 客户端查看(4)的运行结果

最后，@CacheConfig 注解是一个类级别注解，允许类中的方法共享缓存名字，例如：

```
1   @CacheConfig("books")
2   public class BookServiceImpl implements BookService {
3       @Cacheable
4       public Book findBook(String isbn) {...}
5   }
```

至此，已完成 Spring 对 Redis 缓存的整合。

4.3 Redis 缓存优缺点

作为数据交换的缓冲区，缓存的主要作用是提高查询效率。当请求到达服务器时，可以优先查询缓存中的数据，当缓存中存在需要的数据时(缓存命中)，可以直接将缓存中的数据提供给应用程序；当缓存中不存在需要的数据时(缓存未命中)，查询数据库，将查询得到的数据写入缓存并返回给应用程序。图 4-8 演示了带有缓存的存储层架构的设计方案。从 4.2 节给出的范例中可以看出，如果在项目中引入 Redis 缓存，那么在多次发送相同条件的查询请求且缓存命中的情况下，可以直接从 Redis 里获取数据，这会有效地降低数据库的访问压力。使用 Redis 作为缓存有以下优点。

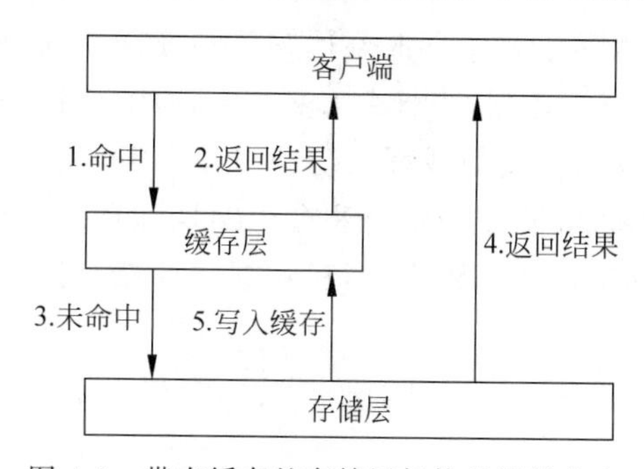

图 4-8 带有缓存的存储层架构的设计方案

(1) 高速读写。因为缓存通常是全内存的(如 Redis、Memcache)，而存储层通常读写性能不够强悍(如 MySQL)，缓存的使用可以有效地加速读写，优化用户体验。

(2) 降低后端负载。缓存可以减少数据库端访问量和复杂计算(例如执行复杂的 SQL 语句)，在很大程度上降低数据库端负载。

同时，使用 Redis 缓存也有一定的代价。

(1) 数据不一致性。缓存层和存储层的数据存在着一定程度的不一致性。这与缓存的数据更新策略有关。

(2) 增加代码维护成本。加入缓存后，需要同时处理缓存层和存储层的数据，增加了开发和维护的成本。

(3) 内存溢出风险。在Java虚拟机管理的Java虚拟机栈、方法区、本地方法栈、程序计数器中，使用堆内缓存可能带来内存溢出风险，从而影响用户进程。

缓存有下述使用场景。

(1) 开销大的复杂计算。以MySQL为例，对于一些复杂的操作或计算（例如大量的表连接操作、分组计算），如果不加缓存，不但无法满足高并发量的要求，也会给MySQL带来巨大的负担。

(2) 加速请求响应。即使查询单条数据足够快，依然可以使用缓存。Redis每秒可以完成数万次读写，并且它提供的批量操作可以优化整个I/O链的响应时间。

在项目开发中常常使用Redis作为缓存。使用缓存还会遇到缓存雪崩和缓存穿透等问题。4.4节和4.5节分别介绍这两个问题及解决方案。

4.4 缓存雪崩

缓存雪崩是指数据未加载到缓存，或者缓存在同一时间大面积失效，导致大量并发请求访问数据库，从而导致数据库主机CPU和内存负载过高，甚至宕机。对于缓存雪崩问题有如下解决方案。

(1) 保持缓存的高可用性。使用Redis的哨兵模式集群或者Redis集群。即使个别Redis节点下线，整个缓存层依然可以使用。

(2) 优化缓存过期时间。使Redis中保存的键永不失效，这样就不会出现大量缓存数据同时失效的问题，但随之而来的是Redis需要更多的存储空间。或者在设计缓存时为每个键选择一个合适的过期时间，将不同键的过期时间哈希开，让缓存失效时间点尽量均匀。

(3) 使用互斥锁。在高并发场景下，为了避免大量的请求同时到达存储层查询数据、重建缓存，可以使用互斥锁控制。如根据键去缓存层查询数据，当缓存未命中时，对键加锁，然后从存储层查询数据，将数据写入缓存层，最后释放锁。其他线程发现获取锁失败，则让线程休眠一段时间后重试。在分布式环境下可以使用Redis的SETNX命令实现互斥锁。分布式环境下使用Redis互斥锁实现缓存重建，优点是设计思路简单，对数据一致性有保障；缺点是代码复杂度增加，会造成大量线程阻塞。

(4) 定时更新策略。对失效性要求不高的缓存，在容器启动时采用定时任务更新或移除缓存。

4.5 缓存穿透

缓存穿透是指查询一个缓存中不存在的数据，由于在Redis缓存中无法查询到需要的数据，这时需要执行数据库查询，如果查询不到数据则不写入缓存，这将导致这个不存在的数据在每次请求到达时都执行数据库查询，进而对数据库产生流量冲击造成缓存穿透。

对于文件4-9中的第26～30行的getUserById()方法，查询id为120这个不存在于数据库的User对象时，每次查询Redis缓存都不可能找到数据，所以会继续向数据库发送查询请求。在此类场景中，由于Redis缓存没有存放数据库里不存在的数据，高并发的查询请求会“穿透”Redis缓存，集中到数据库上。这样会给数据库造成很大的读写压力。一般来讲,解决缓存穿透问题有两种方案：第一,缓存空对象；第二,使用布隆过滤器拦截。

1. 缓存空对象

基本思路：查询数据时,如果缓存层无法命中,仍然将空对象保存到缓存中。这样,大量的查询请求会被 Redis 缓存挡住,不会再执行数据库查询。

缓存空对象同时带来两个问题：第一,缓存了空值意味着缓存层中存储了更多的键,需要更多的内存空间(如果是恶意攻击,则问题更为严重)。比较有效的方法是针对这类数据设置一个较短的过期时间,让其自动从缓存中删除。第二,缓存层和存储层会有一段时间数据不一致。例如,设置过期时间为 5 分钟,如果此时存储层添加了这条数据,此时段就会出现缓存层和存储层数据的不一致,可以利用消息系统或者其他方式清除缓存层中的空对象。下面给出缓存空对象的实现代码,以文件 4-9 中的第 26～30 行的 getUserById()方法为例。

```
1    public User getUserById2(long id) {
2        //在缓存中查询数据
3        User u = template.opsForValue().get("redis_user_" + id);
4        //缓存未命中
5        if(u == null){
6            System.out.println(" ========= DataBase ========= ");
7            //在存储层执行查询
8            u = userDao.getUserById(id);
9            if(u != null)
10               //从存储层获取到数据,将数据写入缓存并设置过期时间为 10 分钟
11               template.opsForValue().setIfAbsent("redis_user_" + id,
12                   u,Duration.ofMillis(10 * 60 * 1000));
13           else
14               //缓存空对象,并设置过期时间为 10 秒
15               template.opsForValue().setIfAbsent("redis_user_" + id,
16                   new User(),Duration.ofSeconds(10));
17       }
18       //返回查询结果
19       if(u!= null&&u.getUserId() == 0)
20           return null;
21       else
22           return u;
23   }
```

2. 使用布隆过滤器拦截

布隆过滤器(Bloom Filter)是 1970 年由布隆提出的。它实际上是一个很长的二进制向量和一系列随机映射函数。布隆过滤器可以用于检索一个元素是否存在于一个集合中。它的优点是空间效率高、查询时间短,缺点是有一定的误识别率和元素删除困难。

布隆过滤器利用位数组很简洁地表示一个集合,并能判断一个元素是否属于这个集合。位数组的每个元素只占用 1 位空间,每个元素只能为 1 或 0。布隆过滤器还拥有 k 个哈希函数。当一个元素加入布隆过滤器时,会使用 k 个哈希函数对其进行 k 次计算,得到 k 个哈希值,并且根据得到的 k 个哈希值,在位数组中把对应位置的值置为 1。判断某个元素是否在布隆过滤器中,就对该元素进行 k 次哈希计算,判断得到的值在位数组中对应位置的值是否都为 1,如果每个元素都为 1,就说明这个元素在布隆过滤器中。因此,如果布隆过滤器判断元素不在集合中,那么肯定不在；如果判断元素在集合中,那么有一定的错误概率。

使用布隆过滤器应对缓存穿透问题的方法如图 4-9 所示。在请求到达缓存层和存储层之前，将存在的键用布隆过滤器提前保存起来。当收到请求时，先在布隆过滤器中查询键是否存在，如果键不存在则直接返回 null，请求不必进入缓存层和存储层执行查询。如果布隆过滤器判断查询的数据存在，则执行缓存层或存储层查询，并返回查询结果。

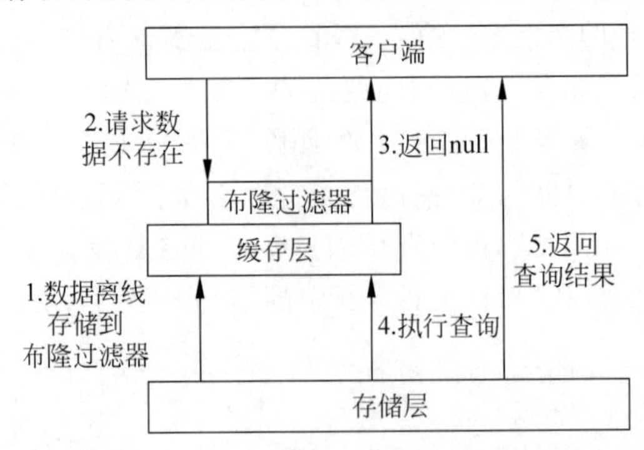

图 4-9 使用布隆过滤器应对缓存穿透问题

下面利用 Google 的 guava 库给出一段布隆过滤器的演示程序。

第一步，引入 guava 的依赖。

```
1  <dependency>
2      <groupId>com.google.guava</groupId>
3      <artifactId>guava</artifactId>
4      <version>30.1-jre</version>
5  </dependency>
```

第二步，编写测试代码。

```
1   @Test
2   public void testBloomFilter() {
3       //元素个数：100万
4       int intersections = 1000000;
5       //创建一个布隆过滤器，初始化大小为100万，误判率为0.01
6       BloomFilter<String> bloomFilter = BloomFilter.create(
7           Funnels.stringFunnel(Charsets.UTF_8), intersections, 0.01);
8       //用于存放所有实际存在的key，判断key是否存在
9       Set<String> set = new HashSet<>(intersections);
10      //用于存放所有实际存在的key，可通过下标取出使用
11      List<String> list = new ArrayList<>(intersections);
12      //加入数据
13      for (int i = 0; i < intersections; i++) {
14          String uuid = UUID.randomUUID().toString();
15          bloomFilter.put(uuid);
16          set.add(uuid);
17          list.add(uuid);
18      }
19      int right = 0;               //正确判断的次数
20      int wrong = 0;               //错误判断的次数
21      for (int i = 0; i < 10000; i++) {
```

```
22              //可以被 100 整除时,取出一个存在的 key,否则随机生成一个 UUID
23              //共取出 100 个存在的 key,随机生成 9900 个 key
24              String data = (i % 100 == 0) ? list.get(i / 100) :
25                  UUID.randomUUID().toString();
26              //判断是否命中
27              if (bloomFilter.mightContain(data)) {
28                  //的确命中
29                  if (set.contains(data)) {
30                      right++;
31                      continue;
32                  }
33                  //判断存在,实际不存在,误判
34                  wrong++;
35              }
36          }
37          //计算命中率和错误率
38          NumberFormat percentFormat = NumberFormat.getPercentInstance();
39          percentFormat.setMaximumFractionDigits(2);
40          float percent = (float) wrong / 9900;
41          float bingo = (float) (9900 - wrong) / 9900;
42          System.out.println("在 100 万个元素中,判断 100 个实际存在的元素,
43              布隆过滤器认为存在的: " + right);
44          System.out.println("在 100 万个元素中,判断 9900 个实际不存在的元素,
45              误认为存在的: " + wrong);
46          System.out.println("命中率: " + percentFormat.format(bingo) +
47              ",误判率: " + percentFormat.format(percent));
48  }
```

运行此段测试代码,控制台输出如图 4-10 所示。

```
在100万个元素中，判断100个实际存在的元素，布隆过滤器认为存在的：100
在100万个元素中，判断9900个实际不存在的元素，误认为存在的：113
命中率：98.86%， 误判率：1.14%
```

图 4-10　布隆过滤器实例代码运行结果

布隆过滤器的用途是判断过滤器中是否存在某数据项,从而减少不必要的数据库查询请求。这种方法适合于数据命中率不高、数据相对稳定、实时性低(通常数据集较大)的应用场景。代码维护较为复杂,缓存空间占用少。表 4-3 总结了前述的缓存穿透问题的两种解决方案。

表 4-3　缓存穿透问题的两种解决方案

缓存穿透的解决方案	适用场景	维护成本
缓存空对象	数据命中率不高 数据频繁变化,实时性高	代码维护简单 需要过多的缓存空间 存在数据不一致问题
布隆过滤器	数据命中率不高 数据相对固定,实时性低	代码维护复杂 缓存空间占用少

4.6 小结

作为内存型 NoSQL 数据库，Redis 特别适合作为缓存。Spring 也提供了相关注解来实现缓存与数据存储层的同步。使用缓存虽然可以提升查询性能，但也存在缓存雪崩和缓存穿透等问题。一般来说，可以通过提高缓存层的可用性防止缓存雪崩，同时利用布隆过滤器及缓存空对象预防缓存穿透。

第 5 章

Redis基础应用

除了可以作为缓存和基于键值对存储的非关系数据库来使用，Redis 还有一些基础应用，如简单的消息队列、可以一次性执行多条命令的流水线及地理位置信息处理等。本章将针对这些基础应用进行介绍。

5.1 发布-订阅

Redis 2.0 及以后的版本支持基于发布-订阅模式的消息机制。在这种模式下，消息发布者和订阅者不直接进行通信。发布者创建一个频道，并在上面发送消息，所有订阅该频道的客户端都能收到消息。Redis 的消息发布-订阅机制如图 5-1 所示，消息订阅者（订阅者 1、订阅者 2 和订阅者 3）订阅频道 Channel 1，消息发布者将消息发布到频道 Channel 1，该消息会被发送给三个订阅者。消息发布者在发送消息后无须关注有多少消息订阅者，更无须关注消息订阅者处理消息的细节，反过来看，消息订阅者也无须关注消息发布者的细节。也就是说，通过消息发布-订阅机制能够最大限度地实现发布者和订阅者间的解耦。同时，可以减少等待消息时不必要的轮询。发布-订阅的应用场景有即时聊天室、公众号订阅等。Redis 适合小型应用系统的发布-订阅，如果是大型应用，还应使用 RabbitMQ 或者 Kafka 等更专业的消息队列软件。

聊天室、公告板可以利用发布-订阅实现消息服务解耦。下面以简单的服务解耦进行说明。如图 5-2 所示，图中有两套业务，上面为朋友圈管理服务，负责管理朋友圈信息；下面为朋友圈频道，用户可以通过各种终端（手机、浏览器等）获取到朋友圈信息。

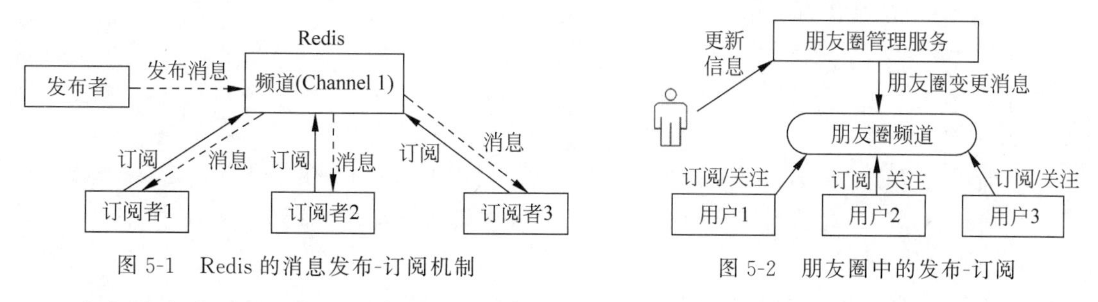

图 5-1 Redis 的消息发布-订阅机制

图 5-2 朋友圈中的发布-订阅

假如发布者更新了朋友圈信息，可以利用发布-订阅模式将更新后的信息发布到指定的频道，其他用户订阅这个频道可以及时更新朋友圈信息，通过这种方式可以有效解决朋友圈管理和朋友圈消息订阅两个业务的耦合性。

5.1.1 常用命令

表5-1列出了Redis提供的一系列用于发布-订阅机制的常用命令。

表5-1 Redis发布-订阅机制的常用命令

常用命令	描 述
PSUBSCRIBE	订阅一个或多个符合指定模式的频道
PUBSUB	查看发布和订阅系统的状态，由多个不同格式的子命令组成
PUBLISH	将消息发送到指定的频道
PUNSUBSCRIBE	退订所有指定模式的频道
SUBSCRIBE	订阅指定的一个或多个频道
UNSUBSCRIBE	退订指定的频道

下面演示发布-订阅是如何工作的。

首先，打开一个客户端连接Redis服务器，作为订阅者接收消息。创建一个订阅频道，命名为redisChat。

```
redis > SUBSCRIBE redisChat
Reading messages... (press Ctrl - C to quit)
1) "subscribe"
2) "redisChat"
3) (integer) 1
```

其次，重新开启一个Redis客户端，作为发布者发送消息，然后在同一个频道redisChat发布两次消息，订阅者就可以接收消息。

```
redis > PUBLISH redisChat "message 1"
(integer) 1
redis > PUBLISH redisChat "message 2"
(integer) 1
```

订阅者的客户端会显示如下消息：

```
redis > SUBSCRIBE redisChat
Reading messages... (press Ctrl - C to quit)
1) "subscribe"
2) "redisChat"
3) (integer) 1
1) "message"
2) "redisChat"
3) "message 1"
1) "message"
2) "redisChat"
3) "message 2"
```

电商平台经常有传送订单数据的需求。下面，利用Jedis的JedisPubSub类来模拟两个模块间通过Redis消息队列交互的动作，来实现订单数据的发送与接收。具体步骤如下。

1. 引入相关依赖

在Maven项目中引入Jedis和GSON的依赖包。

```
1    <dependency>
2        <groupId>redis.clients</groupId>
```

```
3           <artifactId>jedis</artifactId>
4           <version>3.8.0</version>
5       </dependency>
6       <dependency>
7           <groupId>com.google.code.gson</groupId>
8           <artifactId>gson</artifactId>
9           <version>2.8.6</version>
10      </dependency>
```

2. 创建发布者

创建一个名为 Publisher.java 的文件，作为发布者，用于向频道中发送订单数据。代码如文件 5-1 所示。

【文件 5-1】 Publisher.java

```
1   public class Publisher {
2       public static void main(String[] args) {
3           HashMap<String,String> orderHM = new HashMap<String, String>();
4           orderHM.put("id","001");
5           orderHM.put("owner","Peter");
6           orderHM.put("amount","1000");
7           Gson gson = new Gson();
8           String jsonStr = gson.toJson(orderHM);
9           Jedis jedis = new Jedis("localhost",6379);
10          jedis.publish("MQChannel",jsonStr.toString());
11      }
12  }
```

如文件 5-1 所示，第 3～6 行首先创建了 HashMap 类型的订单对象 orderHM，并通过 put()方法模拟实际业务向订单对象中存放数据。由于消息队列一般只传输字符串，在发送前需要将 HashMap 对象转换为 JSON 类型的字符串(第 8 行)。在创建向 localhost:6379 的 Redis 连接后(第 9 行)，通过 Redis 连接对象 jedis 向 MQChannel 频道发送已转换为 JSON 格式的订单数据(第 10 行)。

3. 创建订阅者

创建一个名为 Subscriber.java 的文件，作为订阅者，用于从频道中读取订单数据。代码如文件 5-2 所示。

【文件 5-2】 Subscriber.java

```
1   public class Subscriber extends JedisPubSub {
2       //消息到来时触发
3       public void onMessage(String channel, String message) {
4           Gson gson = new Gson();
5           HashMap<String,String> orderMap =
6                   gson.fromJson(message, HashMap.class);
7           System.out.println(orderMap.get("id"));
8           System.out.println(orderMap.get("owner"));
9           System.out.println(orderMap.get("amount"));
10      }
11
```

```
12        //订阅频道时触发
13        public void onSubscribe(String channel,
14            int subscribedChannels) {
15          System.out.println("subscribe the channel : " + channel);
16        }
17
18        //取消订阅时触发
19        public void onUnsubscribe(String channel,
20            int subscribedChannels) {
21          System.out.println("unsubscribe the channel : " + channel);
22        }
23
24        public static void main(String[] args) {
25          Subscriber subscriber = new Subscriber();
26          Jedis jedis = new Jedis("localhost",6379);
27          jedis.subscribe(subscriber, "MQChannel");
28        }
29    }
```

如文件5-2所示，第1行定义Subscriber类时继承了JedisPubSub类，并分别在第3行、第13行和第19行重写了JedisPubSub类的onMessage()、onSubscribe()和onUnsubscribe()方法。这三个方法分别在收到频道中的消息、订阅频道、取消订阅时触发。

由于在文件5-1中，订单对象以JSON格式发送到频道QMChannel中。因此，在onMessage()方法中会通过第5、6行的代码把接收到的JSON格式的消息转换为HashMap对象，随后第7～9行的输出动作可以被理解为对订单对象的处理。在第24行的main()方法中，在创建Redis连接对象jedis后(第26行)，调用Jedis类的subscribe()方法订阅MQChannel频道(第27行)。注意，该方法的第一个参数subscriber是一个消息处理类，设置后，该对象的onMessage()、onSubscribe()和onUnsubscribe()方法会在特定的时候被触发。

代码完成后，可以首先运行Subscriber.java，可以在控制台看到如下输出。

```
subscribe the channel : MQChannel
```

通过该输出能够确认Subscriber成功地订阅了MQChannel频道。此时，Subscriber还在继续侦听MQChannel频道。接下来可以运行Publisher.java，该段代码没有输出，但是切换到Subscriber.java的控制台，输出内容如图5-3所示。

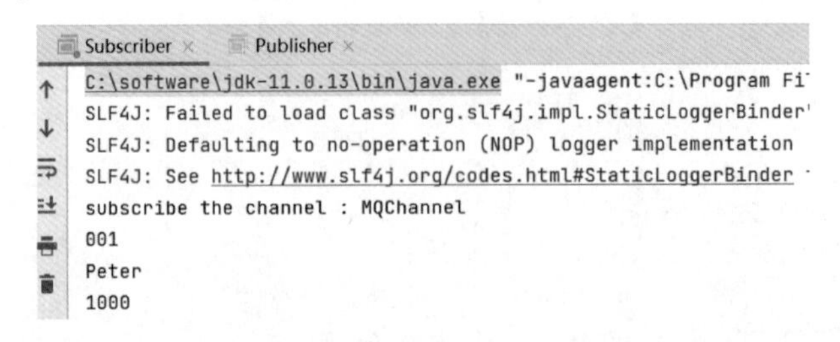

图5-3 发布订单数据后，Subscriber在控制台的输出

其中，第2～4行的输出结果是运行Publisher.java以后得到的，这说明Subscriber.java程序成功地接收到Publisher.java通过MQChannel频道发送过来的订单信息。

5.1.2　消息队列

Spring Data 为 Redis 提供专用的消息传递功能。Redis 消息传递可以分为两个领域：

（1）发布或制作消息。

（2）订阅或消费消息。

对于发布消息，既可以使用 RedisConnection 接口继承自 RedisPubSubCommands 接口（org.springframework.data.redis.connection.RedisPubSubCommands）的 publish()方法，也可以使用 RedisTemplate 类的 convertAndSend()方法。例如：

```
byte[] msg = ...
byte[] channel = ...
con.publish(msg, channel);
```

或

```
RedisTemplate template = ...
template.convertAndSend("hello!", "world");
```

其中，RedisConnection 接口需要原始数据（字节数组）作为参数，而 RedisTemplate 允许将任意对象作为消息传递。

对于订阅消息，RedisConnection 接口提供了 subscribe()和 pSubscribe()方法，分别按通道和模式进行订阅。Spring Data Redis 中的订阅命令是阻塞的。也就是说，调用 subscribe()方法会导致当前线程在开始等待消息时阻塞，只有在取消订阅时才会释放该线程。

对于阻塞式消息订阅，由于要对每个消息侦听器进行连接和线程管理，编码烦琐且性能表现不佳。Spring Data Redis 提供了消息侦听器容器类（RedisMessageListenerContainer），实现异步处理消息。RedisMessageListenerContainer 充当消息侦听器容器。它用于从 Redis 接收消息，并驱动注入其中的消息侦听器（MessageListener）实例。侦听器容器负责管理接收消息的所有线程，并将消息分派到侦听器中进行处理。这使开发人员可以编写与接收消息相关的业务逻辑（并对其做出处理），并将一些 Redis 基础架构问题委托给 Spring 框架。此外，为了使应用程序占用空间最少，RedisMessageListenerContainer 让多个侦听器共享一个连接和一个线程。表 5-2 列举了 Redis 的异步消息处理组件。

表 5-2　Redis 的异步消息处理组件

类 或 接 口	描　　述
org.springframework.data.redis.listener.RedisMessageListenerContainer（类）	为 Redis 消息侦听器提供异步行为的容器。处理侦听、转换和消息调度的细节问题。与底层 Redis（每个订阅一个连接）相反，容器只使用一个连接，该连接对所有注册的侦听器进行“多路复用”，消息调度通过任务执行器完成
org.springframework.data.redis.connection.MessageListener（接口）	消息侦听器
org.springframework.data.redis.listener.adapter.MessageListenerAdapter（类）	消息侦听器适配器，通过反射将消息处理委托给目标侦听器方法，并具有灵活的消息类型转换能力。允许侦听器方法对消息内容进行操作，完全独立于 Redis API

下面介绍两种利用 Spring Data Redis 实现 Redis 发布-订阅的方法：第一种，利用消息侦听器；第二种，利用消息侦听器适配器。下面的例子模拟社交网络中关注公众号一类操

作。具体步骤如下。

1. 配置消息侦听器类

创建一个名为 RedisMessageListener 的消息侦听器类，实现 MessageListener 接口。代码如文件 5-3 所示。

【文件 5-3】 RedisMessageListener. java

```
1   package com.example.redis.message.spring;
2   //import 部分略
3   @Component
4   public class RedisMessageListener implements MessageListener {
5       @Autowired
6       private RedisTemplate template;
7
8       @Override
9       public void onMessage(Message message, byte[ ] pattern) {
10        //获取消息
11        byte[ ] messageBody = message.getBody();
12        //使用值序列化器转换
13        Object msg = template.getValueSerializer()
14              .deserialize(messageBody);
15        //获取侦听的频道
16        byte[ ] channelByte = message.getChannel();
17        //使用字符串序列化器转换
18        Object channel = template.getStringSerializer()
19              .deserialize(channelByte);
20        //频道名称转换
21        String patternStr = new String(pattern);
22        System.out.println(" -- pattern -- :" + patternStr);
23        System.out.println(" -- 频道 -- : " + channel);
24        System.out.println(" -- 消息内容: -- " + msg);
25      }
26  }
```

2. 配置侦听容器并订阅频道

可在文件 3-6 基础上增加消息侦听器容器的配置。修改后的完整代码如文件 5-4 所示。

【文件 5-4】 RedisConfig. java

```
1   package com.example.redis.message.spring;
2   //import 部分略
3   @Configuration
4   @ComponentScan(basePackages = "com.example.redis.message.spring")
5   public class RedisConfig {
6
7       @Bean
8       public RedisConnectionFactory redisConnectionFactory() {
9             //代码见文件 3 - 6
10      }
11      @Bean
12      public RedisTemplate template(@Autowired
13          RedisConnectionFactory rcf){
14          //代码见文件 3 - 6
```

```
15      }
16
17      @Bean
18      public RedisMessageListenerContainer container(
19          RedisConnectionFactory rcf, RedisMessageListener rml) {
20          RedisMessageListenerContainer container =
21              new RedisMessageListenerContainer();
22          container.setConnectionFactory(rcf);
23          container.addMessageListener(rml,
24              new ChannelTopic("redis.life"));
25          container.addMessageListener(rml,
26              new ChannelTopic("redis.news"));
27          /* container.addMessageListener(rml,
28              new PatternTopic("redis.*")); */
29          return container;
30      }
31  }
```

如文件 5-4 所示，在创建消息侦听器容器对象后（第 20～22 行），订阅 redis.life 和 redis.news 两个频道（第 23～26 行）。可以向容器中添加多个消息侦听器。除了按频道名称实现订阅，也可以按模式实现订阅，如第 27、28 行，订阅以 redis.开头的所有频道。

3. 编写测试代码

可以使用 RedisTemplate 的 convertAndSend()方法向指定频道发布消息，同时查看消息处理情况。测试代码如文件 5-5 所示。

【文件 5-5】 TestMessageAsync.java

```
1   //import 部分略
2
3   @RunWith(SpringJUnit4ClassRunner.class)
4   @ContextConfiguration(classes = RedisConfig.class)
5   public class TestMessageAsync {
6       @Autowired
7       private RedisTemplate template;
8
9       @Test
10      public void publish(){
11          template.convertAndSend("redis.life","this is life");
12          template.convertAndSend("redis.news","this is news");
13      }
14  }
```

如文件 5-5 所示，可以利用 RedisTemplate 的 convertAndSend()方法向指定的 redis.life 和 redis.news 频道发送消息。运行测试代码后，控制台输出的消息处理结果如图 5-4 所示。

```
16:58:30.009 [main] DEBUG org.springframework.
--pattern--:redis.news
--pattern--:redis.life
16:58:30.009 [main] DEBUG org.springframework.
16:58:30.009 [main] DEBUG org.springframework.
16:58:30.009 [main] DEBUG org.springframework.
--频道--: redis.news
--频道--: redis.life
--消息内容:--this is news
--消息内容:--this is life
```

图 5-4　消息处理结果

此外，还可以利用消息侦听器适配器实现消息处理。具体步骤如下。

1）定义一个消息接收器类

创建一个名为 MessageReceiver 的消息接收

器类，代码如文件 5-6 所示。

【文件 5-6】 MessageReceiver. java

```
1    @Component
2    public class MessageReceiver {
3        public void receiveMessage(String message, String channel){
4            System.out.println(" -- 频道 -- " + channel);
5            System.out.println(" -- 消息 -- " + message);
6        }
7    }
```

2）配置消息侦听器适配器

可以在文件 5-4 中增加一个消息侦听器适配器 Bean，代码如文件 5-7 所示。

【文件 5-7】 RedisConfig. java 中增加的侦听器适配器 Bean

```
1    @Bean
2    public MessageListenerAdapter listenerAdapter(
3        MessageReceiver receiver) {
4        return new MessageListenerAdapter(receiver, "receiveMessage");
5    }
```

其中，MessageListenerAdapter 的构造方法指定两个参数：第一个参数 receiver 指定消息处理器的代理，即第一步中定义的消息接收器对象；第二个参数指定代理中用来处理消息的方法名，即第一步中 MessageReceiver 类中用来处理消息的方法 receiveMessage()。

3）配置侦听器容器

可以将文件 5-4 中侦听器容器的配置参数由消息侦听器改为消息侦听器适配器，如文件 5-8 所示。

【文件 5-8】 RedisConfig. java 中配置的消息侦听器适配器

```
1     @Bean
2     public RedisMessageListenerContainer container(
3       RedisConnectionFactory factory, MessageListenerAdapter adapter){
4         RedisMessageListenerContainer container = new
5            RedisMessageListenerContainer();
6         container.setConnectionFactory(factory);
7         container.addMessageListener(adapter, new PatternTopic
8            ("redis. * "));
9         return container;
10    }
```

4）运行测试

执行上述修改后，重新运行文件 5-5 的测试代码，可实现同样的消息处理效果。

至此，关于使用 Spring Data Redis API 实现 Redis 发布-订阅应用程序开发已介绍完毕。和很多专业的消息队列系统（例如 Kafka、RabbitMQ）相比，Redis 的发布-订阅略显粗糙，例如无法实现消息堆积和回溯。但 Redis 的发布-订阅足够简单，如果当前场景可以容忍这些缺点，那么使用 Redis 作为消息队列也是一个不错的选择。

5.2 Redis 流

5.2.1 Redis 流简介

Redis 流(Redis Stream)(以下称流)是 Redis 5.0 新增加的数据结构。流的作用类似于只追加(Append-only)日志,主要用于消息队列(Message Queue,MQ)。Redis 有一个发布-订阅机制来实现消息队列的功能,但它有一个缺点就是消息无法持久化。如果出现网络断开、Redis 宕机等情况,消息就会被丢弃。简单来说,发布-订阅可以分发消息,但无法记录历史消息。Redis 流是日志形式的存储结构,可以添加数据。流将为每个数据生成一个时间戳 ID。可以使用这些 ID 检索其关联的条目,或读取和处理流中所有后续条目。因此,流适用于消息队列和时间序列存储。流主要有以下使用场景。

(1) 事件源(例如,跟踪用户操作、点击等)。

(2) 传感器监测(例如,现场设备的读数)。

(3) 通知(例如,在单独的流中存储每个用户的通知记录)。

流的结构如图 5-5 所示。

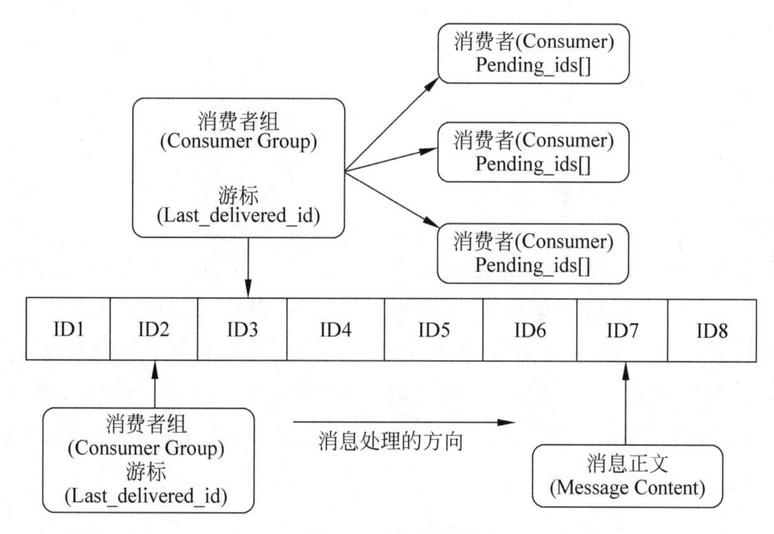

图 5-5 流的结构

如图 5-5 所示,流维护了一个消息链表,将所有加入的消息都串起来,每个消息都有一个唯一的 ID 和对应的内容。其中:

(1) 消息正文(Message Content):存放了每条消息的真实数据。

(2) 消费组(Consumer Group):使用 XGROUP CREATE 命令创建,一个消费组有多个消费者(Consumer)。

(3) 游标(Last_delivered_id):每个消费组会有个游标 Last_delivered_id,任意一个消费者读取了消息都会使游标 Last_delivered_id 往前移动。

(4) 消费者(Consumer):消息的最终消费者,从队列中获取消息进行消费。

(5) 消费者状态变量(Pending_ids):记录了当前已经被客户端读取的消息,但是还没有确认,作用是维护消费者的未确认的 ID。

在介绍流的应用案例前，先了解有关流的几个简单的命令。

(1) XADD：添加消息到末尾。语法：XADD key ID field value[field value...]。注意，如果将 ID 设置为 *，则表示由 Redis 自动生成 ID。

(2) XLEN：获取流包含的元素数量，即消息队列的长度。语法：XLEN key。

(3) XRANGE：获取消息列表，会自动过滤已经删除的消息。语法：XRANGE key start end[COUNT count]。其中，start 为"－"表示最小值，end 为"＋"表示最大值。COUNT 选项可指定返回的条目数。

下面列举一个应用流命令的简单示例。

(1) 向流中添加一条消息。

```
redis > XADD mystream * name LiuZongYuan age 16
"1674825991997 - 0"
```

(2) 获取 key 为 mystream 的流的长度。

```
redis > XLEN mystream
(integer) 1
```

(3) 获取 key 为 mystream 的流中的所有消息。

```
redis > XRANGE mystream - +
1) 1) "1674825991997 - 0"
   2) 1) "name"
      2) "LiuZongYuan"
      3) "age"
      4) "16"
```

Redis 流操作可分为两个功能：追加记录与消费记录。这种模式与发布-订阅有相似之处，其主要区别在于消息的持久性以及消息的使用方式。发布-订阅依赖于瞬时消息的广播(即，如果你不听，你就会错过一条消息)，但流使用一种持久的、只追加的数据类型，可以在流被修剪之前保留消息。消息使用方式的区别是发布-订阅注册了服务器端订阅，Redis 将到达的消息推送到客户端，而流需要主动轮询。在 Spring Data Redis 中，流的主要操作由 org. springframework. data. redis. connection 包和 org. springframework. data. redis. stream 包中的类或接口提供。

5.2.2 Redis 流操作之追加

在 Spring Data Redis 中，如果要向流追加记录，与其他操作一样，可以使用 RedisConnection 接口或 StreamOperations < K，HK，HV >接口(org. springframework. data. redis. core. StreamOperations < K，HK，HV >)来实现。这两个接口分别提供了 xAdd() 和 add()方法用于向流中追加记录。其中 RedisConnection 接口提供的 xAdd()方法以原始数据(字节数组)作为参数，而 StreamOperations 接口提供的 add()方法允许将任意对象作为记录传入。

1. 使用 RedisConnection 接口执行追加

下面的例子模拟利用流存储传感器传送的实时数据，如温度、压力等。创建一个名为 StreamRedisConfig. java 的类。代码如文件 5-9 所示。

【文件 5-9】 StreamRedisConfig.java

```
1    @Configuration
2    public class StreamRedisConfig {
3        @Bean
4        public RedisConnectionFactory redisConnectionFactory() {
5            //代码可见文件 3-6
6        }
7    }
```

如文件 5-9 所示，在 StreamRedisConfig 配置类中创建 RedisConnectionFactory 的配置元数据，代码可参考文件 3-6。随后，编写测试代码，利用 RedisConnection 向流中追加记录，测试代码如文件 5-10 所示。

【文件 5-10】 TestStreamsDemo.java

```
1    @RunWith(SpringJUnit4ClassRunner.class)
2    @ContextConfiguration(classes = StreamRedisConfig.class)
3    public class TestStreamsDemo {
4        @Autowired
5        private RedisConnectionFactory rcf;
6
7        @Test
8        public void testAppendByConnection() {
9            RedisConnection conn = rcf.getConnection();
10           byte[] stream = "my-stream".getBytes();
11           Map<byte[],byte[]> map = new HashMap<>();
12           map.put("sensor".getBytes(),"tp-sensor".getBytes());
13           map.put("temperature".getBytes(),"21.0C".getBytes());
14           map.put("pressure".getBytes(),"3500Kpa".getBytes());
15           ByteRecord record =
16               StreamRecords.rawBytes(map).withStreamKey(stream);
17           conn.xAdd(record);
18       }
19   }
```

如文件 5-10 所示，第 9 行利用 RedisConnectionFactory 对象获取 RedisConnection 对象。由于 RedisConnection 只能接受字节数组作为参数，第 10～14 行将追加到流中的键（第 10 行）、字段和值（第 12～14 行）都转换为字节数组。第 15、16 行首先调用 StreamRecords 类的静态方法 rawBytes()，为给定的原始字段值对创建新的 ByteRecord，其中 ByteRecord 接口代表流中由二进制字段值对集合所支持的记录。随后，调用 ByteRecord 对象的 withStreamKey(byte[]key)方法，创建一个新的指定了键的 ByteRecord 对象。第 17 行调用 RedisConnection 接口继承自 RedisStreamCommands 接口的 xAdd()方法。xAdd()方法的原型如下：

```
@Nullable
default RecordId xAdd(byte[] key, Map<byte[],byte[]> content)
```

该方法的功能是向键 key 指定的流中追加一条新的记录，记录的内容由 Map 对象中的字段和值组成。运行此测试代码后，可以在 Redis 的客户端利用 XRANGE 命令查看运行结果：

```
redis > XRANGE my-stream - +
1) 1) "1674890806976-0"
   2) 1) "temperature"
      2) "21.0C"
      3) "sensor"
      4) "tp-sensor"
      5) "pressure"
      6) "3500Kpa"
```

2. 使用 StreamOperations 接口执行追加

对于文件 5-10 中的测试数据,也可以利用 StreamOperations 接口提供的 add()方法追加到流。而 StreamOperations 对象需要调用 RedisTemplate 类的 opsForStreams()方法获取。因此,可在文件 5-9 的基础上增加 RedisTemplate 类的配置元数据,代码如文件 5-11 所示。

【文件 5-11】 文件 5-9 增加的配置元数据

```
1    @Bean
2    public RedisTemplate<String, Map<String,String>> template(
3      @Autowired RedisConnectionFactory rcf){
4      RedisTemplate<String, Map<String,String>> template = new
5        RedisTemplate<>();
6      template.setConnectionFactory(rcf);
7      template.setDefaultSerializer(RedisSerializer.string());
8      return template;
9    }
```

Spring Data Redis 中,所有发送到流的记录都需要序列化为二进制格式。因此,文件 5-11 的第 7 行将 Redis 默认的序列化器设置为 StringRedisSerializer。

随后,可以编写测试代码,利用 StreamOperations 对象完成追加。可在文件 5-10 的基础上增加一个测试用例,如文件 5-12 所示。

【文件 5-12】 文件 5-10 增加的测试用例

```
1    @Test
2    public void testAppendByTemplate() {
3      Map<String,String> map = new HashMap<>();
4      map.put("sensor","tp-sensor2");
5      map.put("temperature","62.0C");
6      map.put("pressure","4300Kpa");
7      StreamOperations<String,String,String> ops =
8        template.opsForStream();
9      ops.add("my-stream2",map);
10   }
```

5.2.3 Redis 流操作之消费

消费者可以消费一个或多个流。流提供读取命令,允许从已知流中的任意位置消费流(随机访问),并且可以超出流末端消费新的流记录。RedisConnection 接口从 RedisStreamCommands 接口继承了 xRead()和 xReadGroup()方法,分别对应 Redis 命令 XREAD 和 XREADGROUP,

从一个或多个流中读取数据以及在消费者组内进行读取。Redis中的订阅命令可能会被阻塞。也就是说，在Redis连接上调用xRead()方法会导致当前线程在开始等待消息时阻塞。只有当读取命令超时或收到消息时，线程才会被释放。要使用流消息，可以轮询应用程序代码中的消息，或者通过消息侦听器容器使用两种异步接收方式中的一种，即命令式或响应式。每次新记录到达时，容器都会通知应用程序代码。接下来，本节从同步接收、异步接收和消息偏移量三方面介绍流的消费操作。

1. 同步接收消息

虽然流消费通常与异步处理相关，但也可以从流中同步消费消息。StreamOperations接口中重载的read()方法可以实现这个功能。在同步接收期间，调用线程可能会阻塞，直到消息可用为止。例如，准备消费流中已追加的数据，可以利用文件5-13所示的代码消费流数据（流中的数据已在文件5-10和文件5-12中完成追加）。

【文件5-13】　部分消费者代码

```
1   @Test
2   public void testRecMessageSynchronously() {
3     StreamOperations<String,String,String> ops =
4         template.opsForStream();
5     List<MapRecord<String, String, String>> messages = ops.read(
6       StreamReadOptions.empty(),StreamOffset.fromStart("my-stream1"));
7     if(messages!= null)
8         messages.forEach(msg -> System.out.println(msg.getStream() + ":" +
9         msg.getId() + " : " + msg.getValue()));
10  }
```

如文件5-13所示，第3、4行获取StreamOperations对象。第5行调用StreamOperations接口的read()方法消费（读取）流中的数据。read()方法中指定了两个参数：第一个指定读操作的参数，通过调用StreamReadOptions类的静态方法empty()创建一个空的StreamReadOptions对象；第二个参数指定要读取的流。第6行指定从my-stream1流的起始位置开始读取。对于读取到的流中的记录（MapRecord）列表，利用Lambda表达式对列表中的每个元素逐一处理（第8、9行）。其中的记录都是由字段值对（Field-Value Pairs）组成的。MapRecord接口继承了Record接口的三个方法，其中getStream()方法返回流在Redis中的键；getId()方法返回流中实体的ID；getValue()方法返回流中实体的值。此测试用例的运行结果如图5-6所示。

```
my-stream1 : 1674913205847-0 : {sensor=tp-sensor1, temperature=21.0C, pressure=3500Kpa}
```

图5-6　文件5-13测试用例运行结果

在某些应用场景中，需要的不是向许多客户端提供相同的消息流，而是向客户端提供同一消息流的不同子集。例如，对于一个处理速度慢的消息，可以让N个不同的工作人员接收消息流的不同部分，这就需要将不同的消息路由到不同的工作人员来处理消息。为了实现这一点，Redis使用了一个叫作消费者组的概念。从实现的角度来看，Redis消费者组与Kafka消费者组无关，尽管它们在功能上是相似的。消费者组就像一个从流中获取数据的伪消费者，实际上为多个消费者提供服务，并做出如下保证。

（1）每个消息都提供给不同的消费者，不可能将同一消息传递给多个消费者。

（2）在消费者组中，每个消费者需要通过名称来标识。名称是区分大小写的字符串，每个消费者必须指定名称。

（3）每个消费者组都具有一个“从未使用过的第一个 ID”的概念，其作用相当于书签。因此，当消费者请求新消息时，流可以只提供以前未传递的消息。

（4）使用（或消费）消息需要使用 XACK 命令进行确认。Redis 将确认解释为：此消息已正确处理，可以将其从消费者组中移除。

（5）消费者组会跟踪当前挂起的所有消息，即已传递给消费者组中的某个消费者但尚未确认为已处理的消息。当查看流消息的历史记录时，每个消费者只能看到传递给它自己的消息。

要使用消费者组处理消息，首先要在 Redis 中创建消费者组。可以使用下述命令创建消费者组：

```
XGROUP CREATE mystream mygroup $
```

其中，mystream 表示已创建流的键，mygroup 表示指定的消费者组的名称，$ 是消费者组的 ID。$ 的含义为只有到达流中的新消息才会提供给组中的消费者。如果将 ID 指定为 0，则消费者组将使用流历史记录中的所有消息。当然，也可以将 ID 指定为其他的合法值。可利用文件 5-10 或文件 5-12 创建的流创建消费者组 myGroup，并指定 myGroup 消费流中的所有消息，编写的测试代码如文件 5-14 所示。

【文件 5-14】 消费者组测试代码

```
1   @Test
2   public void testRecMessageSynchronouslyByGroup() {
3     StreamOperations<String, String, String> ops =
4        template.opsForStream();
5     List<MapRecord<String, String, String>> messages = ops.read(
6        Consumer.from("myGroup","myConsumer"),
7        StreamReadOptions.empty(),
8        StreamOffset.create("my-stream1",ReadOffset.lastConsumed()));
9     if(messages!= null){
10      messages.forEach(msg -> { System.out.println(
11        msg.getStream() + " : " + msg.getId() + " : " + msg.getValue());
12        ops.acknowledge("myGroup",msg);
13    });}
14  }
```

如文件 5-14 所示，第 5 行调用 StreamOperations 接口的 read()方法消费（读取）流数据，需要指定 read()方法的三个参数（第 6～8 行）。第 6 行调用 Consumer 类的静态方法 from()创建一个新的消费者，指定消费者组为 myGroup、消费者名称为 myConsumer。第 8 行指定从 my-stream1 流中消费消息。第 10、11 行用于消费消息。通过消费组读取消息时，服务器将记录给定的消息已发送，并将其添加到已发送但尚未确认的消息列表（Pending Entries List，PEL）中。第 12 行调用 StreamOperations 接口的 acknowledge()方法向 Redis 服务器确认当前消息已被处理，可以将其从消费者组中移除。

运行文件 5-14 的测试代码，控制台的输出如图 5-6 所示。此外，还可以通过 RedisInsight 查看第 12 行发送确认命令（XACK 命令）前后 Redis 消息队列的变化情况。图 5-7 和图 5-8

分别展示了发送确认命令前后的情况。

Group ... ↑	Consumers	Pending
myGroup	1	1

图 5-7　发送确认命令前的情况

Group ... ↑	Consumers	Pending
myGroup	1	0

图 5-8　发送确认命令后的情况

如图 5-7 所示，如果不发送确认命令(移除文件 5-14 第 12 行代码)，Redis 服务器显示有一条消息挂起(Pending)。当执行确认命令后，则该消息从流中移除，挂起的消息数量为 0，如图 5-8 所示。

2. 异步接收消息

由于其阻塞特性，同步接收消息没有吸引力，因为它需要对每个消费者进行连接和线程管理。为了解决这个问题，Spring Data 提供了消息侦听器，它可以实现异步接收消息。Spring Data 提供了两种针对不同编程模型的消息侦听器容器。

(1) StreamMessageListenerContainer 接口充当命令式编程模型的消息侦听器容器。它用于消费来自流的记录，并驱动注入其中的 StreamListener 实例。

(2) StreamReceiver 接口提供了消息侦听器的响应式变体。它可以将来自流的消息作为潜在的无限流使用，并通过 Flux(一个利用单向数据流实现的应用架构)发出流消息。

StreamMessageListenerContainer 和 StreamReceiver 负责将接收消息和发送消息的所有线程交由侦听器处理。消息侦听器容器(或消息接收器)是消息驱动 POJO(Message Driven POJOs，MDPs)和消息提供者之间的中介，负责注册以接收消息、获取和释放资源、执行异常转换等。这使得应用程序开发人员可以编写与接收消息相关的业务逻辑，其余细节问题交给框架处理。

这两个容器都允许更改运行时配置，以便在应用程序运行时添加或删除订阅。此外，容器使用延迟订阅方法，仅在需要时使用 RedisConnection。如果所有侦听器都被取消订阅，那么容器会自动执行清理，并释放线程。

3. 消息偏移量

执行流读取操作时，Redis 从给定的偏移量开始读取流记录。ReadOffset 类(org.springframework.data.redis.connection.stream.ReadOffset)定义了读取偏移量(文件 5-14 第 8 行)。Redis 支持 3 种偏移量，具体取决于是独立使用流还是在消费者组内使用流。

(1) ReadOffset.latest()：读取最新消息。

(2) ReadOffset.from()：读取特定消息 ID 之后的消息。

(3) ReadOffset.lastConsumed()：读取最后使用的消息 ID 之后的消息(仅针对消费者组)。

5.2.4　Redis 流操作之序列化

所有发送到流中的记录都需要序列化。由于流与哈希数据结构非常接近，因此流的键、字段和值均使用了 RedisTemplate 配置的序列化器。下面列出了相关的序列化器。

(1) 键使用序列化器 keySerializer，用于 Record#getStream()方法。

（2）字段使用序列化器 hashKeySerializer，用于序列化哈希中的每个字段。

（3）值使用序列化器 hashValueSerializer，用于序列化哈希中的每个值。

如果不使用任何序列化程序，则需要确保记录值已经是二进制的。对于 5.2.2 节和 5.2.3 节的例子，已在文件 5-11 中设置使用默认的序列化器 StringRedisSerializer。

向流中添加一个复杂值可以通过三种方式完成。

（1）将复杂值转换为简单值，如使用 JSON 字符串表示。

（2）使用合适的序列化器序列化值。

（3）使用 HashMapper 将值转换为适合序列化的 Map。

其中，第一种方式是最直接的，但忽略了流结构所提供的字段值，流中的值对其他消费者来说仍然是可读的。第二种方式的优点与第一种方式相同，但要求所有的消费者都必须实现相似的序列化机制。第三种方式，使用 HashMapper 接口是一种更复杂的方法，它使用了流的哈希结构，但对源代码进行了扁平化处理。只要选择了合适的序列化程序，其他用户仍然能够读取记录。

HashMapper 接口可将 Java 对象转换为 Redis 的哈希类型或 Map 类型(java.util.Map)。需要使用能够序列化和反序列化的哈希键和哈希值序列化器。如，要将一个普通的 User 对象发送到流中，可按以下步骤执行。

第一步，创建 User 类，代码如文件 5-15 所示。

【文件 5-15】 User.java

```
1    @Data
2    public class User {
3      private int id;
4      private String name;
5      private String pass;
6      public User(int id, String name, String pass) {
7         this.id = id;
8         this.name = name;
9         this.pass = pass;
10     }
11     //此处省略了 toString()方法
12   }
```

第二步，编写测试代码，向流中发送 User 对象，如文件 5-16 所示。

【文件 5-16】 部分测试代码

```
1    @Test
2    public void testStreamForComplexValue() {
3      StreamOperations<String,String,User> ops = template.opsForStream();
4      ObjectRecord<String, User> record = StreamRecords.newRecord()
5         .in("user-stream").ofObject(new User(1,"Tom","angel"));
6      ops.add(record);
7      List<ObjectRecord<String,User>> records = ops.read(
8         User.class,StreamOffset.fromStart("user-stream"));
9      records.forEach(rec->System.out.println("id=" + rec.getId() +
10     ",stream=" + rec.getStream() + ",value=" + rec.getValue().toString()));
11   }
```

如文件 5-16 所示，第 4 行调用 StreamRecords 类的 newRecord()方法创建记录构建器(org.springframework.data.redis.connection.stream.StreamRecords.RecordBuilder < S >)用以在流中创建记录。随后，调用 RecordBuilder 类的 in()方法来配置流的键。再调用 RecordBuilder 类的 ofObject()方法创建一个 Object 类型的记录(第 5 行)。第 6 行调用 add()方法向流中发送 User 对象。第 8、9 行从流中取出记录并消费(第 9、10 行)。运行测试代码后的控制台输出如图 5-9 所示。

```
id =1674999570815-0, stream=user-stream,value= User{id=1, name='Tom', pass='angel'}
```

图 5-9　控制台输出

5.3　流水线

Redis 流水线(Pipelining)技术是一种提高 Redis 性能的技术，它允许同时发出多个命令而无须等待每个命令的单独响应，可以在全部命令执行结束后一次性返回运行结果。Redis 是一个使用客户-服务器(C/S)模型的 TCP 服务器，它采用与 HTTP 类似的请求-响应协议。一般情况下，用户每执行一个 Redis 命令，客户端与服务器都需要进行一次通信。客户端将命令发送给服务器，随后客户端会阻塞并等待 Redis 服务器处理，服务器将命令执行的结果以响应报文的形式返回给客户端。执行一条 Redis 命令分为如下 4 个步骤：

(1) 发送命令。

(2) 命令排队。

(3) 执行命令。

(4) 返回结果。

这 4 个步骤的时间消耗称为往返时间(Round Trip Time，RTT)。大部分的 Redis 命令是不支持批量操作的。例如，要执行 n 次 HGETALL 命令，需要消耗 n 次往返时间。实际项目中，Redis 的客户端和服务器可能部署在不同的机器上。例如，客户端在 A 地，Redis 服务器在 B 地，两地直线距离约为 2600km，那么 1 次 RTT 时间=2600×2/(300 000×2/3)=26ms(光在真空中的传播速度为每秒 3×10^8 m，这里假设光在光纤中的传播速度为真空光速的 2/3)，那么客户端在 1 秒内大约能执行 40 条命令，这与 Redis 的高并发高吞吐特性背道而驰。

Redis 的流水线机制能够改善上述问题。流水线能够将一组 Redis 命令进行组装，一次向服务器发送多条命令而无须等待回复，然后在一个步骤中读取回复。在没有使用流水线时执行 n 条命令的情形如图 5-10 所示，使用流水线时执行 n 条命令的情形如图 5-11 所示。

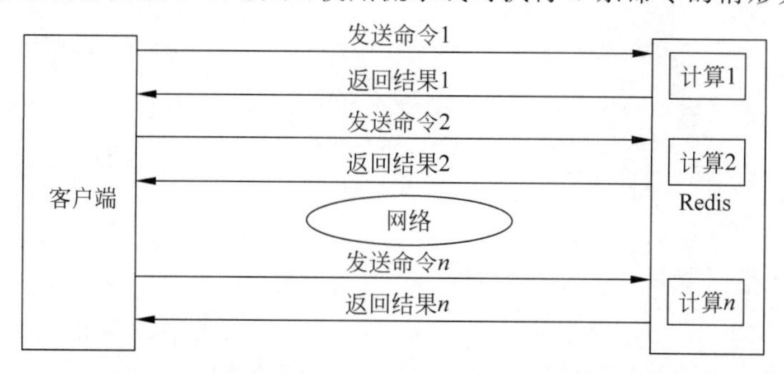

图 5-10　没有使用流水线时执行 n 条命令的情形

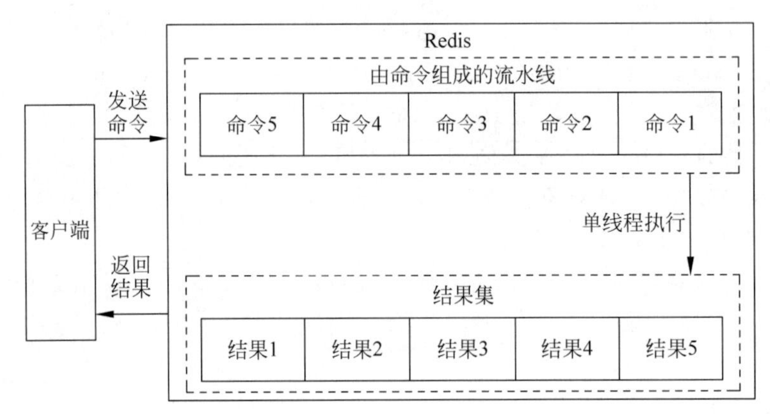

图 5-11　使用流水线时执行 n 条命令的情形

在没有使用流水线的情况下，执行 n 条命令耗时 $n*$ RTT；而使用流水线执行 n 条命令时，耗时 RTT。流水线方式可以大量节省命令和结果的传输时间，从而提高程序性能。

可以使用 RedisTemplate 来执行流水线操作，RedisTemplate 提供的流水线方法如表 5-3 所示。

表 5-3　RedisTemplate 提供的流水线方法

方法原型	说明
List<Object> executePipelined(RedisCallback<?> action)	在流水线上执行给定的操作 action 并返回结果
List<Object> executePipelined(RedisCallback<?> action, @Nullable RedisSerializer<?> resultSerializer)	在流水线上执行给定的操作 action 并使用专用的序列化器 resultSerializer 返回结果
List<Object> executePipelined(SessionCallback<?> session)	在流水线上执行 Redis 会话 session
List<Object> executePipelined(SessionCallback<?> session, @Nullable RedisSerializer<?> resultSerializer)	在流水线上执行 Redis 会话 session 并使用专用的序列化器 resultSerializer 返回结果

在下面的范例中，向 Redis 中批量插入 10 000 个键，比较了不采用流水线方式和采用流水线方式时的性能差异。代码如文件 5-17 所示。

【文件 5-17】 TestPipelineDemo.java

```
1   @RunWith(SpringJUnit4ClassRunner.class)
2   @ContextConfiguration(classes = RedisConfig.class)
3   public class TestPipelineDemo {
4
5       @Autowired
6       private StringRedisTemplate template;
7
8       private final int batchSize = 10000;
9       @Test
10      public void testPipeline() {
11          long start = System.currentTimeMillis();
12          for(int cnt = 0; cnt < batchSize;cnt++){
13              template.opsForValue().set("key" + cnt, String.valueOf(cnt));
14              template.opsForValue().get("key" + cnt);
```

```
15          }
16          long end = System.currentTimeMillis();
17          System.out.println("time cost by Non-pipeline :"
18               + (end-start) + " ms.");
19
20          long s = System.currentTimeMillis();
21          template.executePipelined((RedisCallback<Object>) conn -> {
22            for(int i = 0; i < batchSize; i++) {
23           conn.set(("key" + i).getBytes(),String.valueOf(i)
24               .getBytes());
25           conn.gct(("key" | i).gelBytes());
26          }
27          return null;
28          });
29          long e = System.currentTimeMillis();
30          System.out.println("time cost by pipeline :" + (e-s) + " ms.");
31      }
32  }
```

其中,文件 5-17 中的配置类 RedisConfig(第 2 行)的内容见文件 3-6,此处不再赘述。在获取 StringRedisTemplate 的实例后(第 5、6 行),执行了 10 000 次读写操作(非流水线作业),并在第 17、18 行输出这 10 000 次操作所耗费的时间。

随后,在第 21 行调用 RedisTemplate 类的 executePipelined()方法实现流水线作业。该方法的用法与表 3-7 中的 execute()方法用法类似,可以接收的参数为 RedisCallback 对象或 SessionCallback 对象(第 21 行)。第 22～26 行执行 10 000 次读写操作(流水线作业)。由于操作的是字符串,RedisTemplate 会自动用值序列化器在返回响应数据前反序列化所有结果。从 RedisCallback 对象返回的值必须为 null(第 27 行),因为该值会被丢弃,从而支持返回流水线命令的执行结果。

上述测试代码的运行结果如图 5-12 所示。可见非流水线方式执行了多次 Redis 连接的建立与关闭,程序运行耗时 3137ms,而流水线方式只执行了一次连接建立与关闭,程序运行耗时 264ms。

```
16:43:11.202 [main] DEBUG org.springfr
16:43:11.202 [main] DEBUG org.springfr
16:43:11.202 [main] DEBUG org.springfr
16:43:11.202 [main] DEBUG org.springfr
16:43:11.202 [main] DEBUG org.springfr
time cost by Non-pipeline :3137 ms.
16:43:11.210 [main] DEBUG org.springfr
16:43:11.466 [main] DEBUG org.springfr
time cost by pipeline :264 ms.
```

图 5-12　文件 5-17 测试代码运行结果

目前,大多数的 Redis 客户端已支持流水线操作。流水线操作可以在一定程度上提升程序性能,但是每次由流水线组装的命令个数不能没有限制。一次组装在流水线中的数据量过大,一方面会增加客户端的等待时间,另一方面会造成一定程度的网络拥塞。可以将一个包含大量命令的流水线拆分成多个较小的流水线来完成。同时,流水线只能操作一个 Redis 实例。即使在分布式 Redis 场景中,也可以作为批量操作的重要优化手段。

5.4 事务与 Lua

Redis 事务是一组命令的集合,事务支持一次执行多条命令。Redis 事务具有以下特征:事务中的所有命令都会按顺序执行,其他客户端提交的命令不会插入事务执行的命令序列中,并且事务中所有命令都会被序列化;这就保证了事务中的一组命令成为一个独立的执行单元。即 Redis 事务就是一次性、顺序性、排他性地执行一个队列中的一组命令。简单地说，事务表示一组动作要么全部执行，要么全部不执行。例如，在社交网站上用户 A 关注了用户 B，那么需要在用户 A 的关注列表中加入用户 B，并且在用户 B 的关注者列表中添加用户 A，这两个动作要么全部执行，要么全部不执行，否则会出现数据不一致的情况。

Redis 事务没有隔离级别的概念，也不保证事务的原子性。为保证多条命令组合的原子性，Redis 借助 Lua 脚本来解决这个问题。本节首先介绍 Redis 事务的使用方法和它的局限性，之后介绍 Lua 脚本的基本使用方法，以及如何将 Redis 和 Lua 脚本集成，最后给出 Spring Data Redis 操作 Lua 的相关例子。

5.4.1 Redis 事务

Redis 提供了简单的事务功能，表 5-4 列出了 Redis 事务的常用命令及描述。

表 5-4 Redis 事务的常用命令及描述

命令	描　述
DISCARD	取消事务，取消执行事务块内的所有命令
EXEC	执行所有事务块内的所有命令
MULTI	标记一个事务块的开始
UNWATCH	取消 WATCH 命令对所有键的监视
WATCH	监视所有的键，如果在事务执行之前这些键被其他命令所改动，那么事务将被打断

利用 Redis 处理事务时，将一组需要一起执行的命令放到 MULTI 和 EXEC 两个命令之间。这两个命令之间的命令是按顺序执行的原子操作。例如，下面的操作实现了上述的用户关注问题。

```
redis> MULTI
OK
redis> SADD user:a:follow user:b
QUEUED
redis> SADD user:b:fans user:a
QUEUED
```

此时，可以看到 SADD 命令的返回结果是 QUEUED，代表命令并没有真正执行，而是暂时保存在 Redis 中。此时，如果另一个客户端执行命令 SISMEMBER user:a:follow user:b 则返回结果应该是 0。

```
redis> SISMEMBER user:a:follow user:b
(integer) 0
```

只有当 EXEC 命令执行后，用户 A 关注用户 B 的行为才算完成。如下命令序列所示。

```
redis> MULTI
OK
redis> SADD user:a:follow user:b
QUEUED
redis> SADD user:b:fans user:a
QUEUED
redis> EXEC
1) (integer) 1
2) (integer) 1
```

```
redis > SISMEMBER user:a:follow user:b
(integer) 1
```

如果要停止事务的执行，要使用 DISCARD 命令代替 EXEC 命令。

```
redis > DISCARD
OK
redis > SISMEMBER user:a:follow user:b
(integer) 0
```

有些应用场景需要在事务执行前，确保事务中的键没有被其他客户端修改过才能执行事务，否则不执行(类似乐观锁)。Redis 提供了 WATCH 命令来解决这类问题，表 5-5 展示了事务中 WATCH 命令执行时序。

表 5-5 事务中 WATCH 命令执行时序

时 间 点	客户端 1	客户端 2
T1	SET num 10	
T2		WATCH num
T3	WATCH num	
T4	MULTI	
T5		MULTI
T6		INCRBY num 5
T7	INCRBY num 8	
T8		EXEC (success)
T9	EXEC (nil)	

如表 5-5 所示，客户端 1 和客户端 2 均在执行 MULTI 命令之前执行了 WATCH 命令，用于监控 num 的值。T6 时刻，客户端 2 修改了 num 的值。客户端 1 在 T7 时刻也修改了 num 的值。但在 T8 时刻，客户端 2 执行 EXEC 命令，完成对 num 值的修改。而 T9 时刻，客户端 1 执行 EXEC 命令，此时，客户端 1 对 num 值的修改无效，事务回滚(EXEC 命令执行结果为 nil)。WATCH 可以同时监控多个键，在监控期间只要有一个键被其他客户端修改，则整个事务回滚。因此，这番操作后，num 的值应为 15。并且，WATCH 命令的生命周期只和事务关联，一个事务执行完毕，相应的 WATCH 命令的生命周期就会结束。

Redis 对事务的支持是很弱的，也无法实现命令之间的逻辑关系计算。5.4.2 节要介绍的 Lua 脚本同样可以实现事务的相关功能，但功能要强大很多。

5.4.2 Lua 脚本

Lua 是一种轻量级的脚本语言，是巴西一所大学的研究小组发明的，其设计目的是作为嵌入式应用程序移植到其他应用。结合 Redis 和 Lua 脚本语言的特性，如果在 Redis 应用中遇到如下需求，就可以引入 Lua 脚本：

(1) 重复执行相同类型的命令。如要在内存中缓存数字 1～1000。

(2) 在高并发场景下减少网络开销。如要向 Redis 服务器发送多条命令，就可以把这些命令放入 Lua 脚本里，这样通过调用脚本只耗费一次网络开销就能执行多条 Redis 命令。

(3) Redis 会把 Lua 脚本作为一个整体来执行，天然具有原子性。所以，在一些需要原子操作的场景里，Lua 脚本非常适合。

作为编程语言，Lua本身就包含了变量、操作符、循环等语法。本节只给出与Redis应用相关的Lua语法。

1. 数据类型及逻辑处理

Lua提供了如下几种数据类型：booleans（布尔）、numbers（数值）、strings（字符串）、tables（表格）。下面结合例子对Lua的基本数据类型和逻辑处理进行说明。

（1）字符串。

下面定义一个字符串类型的数据：

```
1    local strings s = "world"
```

其中，local代表s是一个局部变量，如果没有local，则代表全局变量。print()函数可以输出变量的值，例如下面代码将输出world，其中--是Lua的注释。

```
1    -- 输出 world
2    print(s)
```

（2）数组。

在Lua中，如果要使用类似数组的功能，可以用tables类型。下面代码定义了一个tables类型的变量myArr，但和大多数编程语言不同的是，Lua数组的下标从1开始：

```
1    local tables myArr = {"redis", "jedis", true, 80.0}
2    -- 输出 true
3    print(myArr[3])
```

如果要遍历整个数组，可以使用for和while，这些关键字和许多编程语言是一致的。

① for。

下面代码会计算1～100的和，循环体以end作为结束符：

```
1    local int sum = 0
2    for i = 1, 100
3    do
4        sum = sum + i
5    end
6    -- 输出结果为 5050
7    print(sum)
```

要遍历myArr数组，首先要知道tables的长度，只需要在变量前加一个#即可：

```
1    for i = 1, #myArr
2    do
3        print(myArr[i])
4    end
```

除此之外，Lua还提供了内置函数ipairs()，使用for index，value in ipairs(tables)可以遍历所有的索引下标和值：

```
1    for index, value in ipairs(myArr)
2      do
3          print(index)
```

```
4           print(value)
5     end
```

② while。

下面的代码同样会计算 1～100 的和。

```
1     local int sum = 0
2     local i = 0
3     while i <= 100
4     do
5         sum = sum + i
6         i = i+1
7     end
8     --结果为 5050
9     print(sum)
```

③ if else。

要确定数组 myArr 中是否包含 jedis，可以使用下面的 Lua 代码：

```
1     local tables myArr = {"redis","jedis",true,80.9}
2     for i = 1, #myArr
3     do
4         if myArr[i] == "jedis"
5         then
6             print("true")
7             break
8         else
9             -- do nothing
10        end
11    end
```

(3) 哈希。

如果要使用类似哈希的功能，同样可以使用 tables 类型，例如下面代码定义了一个 tables，每个元素包含了键和值，其中运算符..的作用是连接两个字符串。

```
1     local tables user_1 = {age = 28, name = "Tome"}
2     print("user_1 ages is "..user_1["age"])
```

如果要遍历 user_1，可以使用 Lua 的内置函数 pairs()：

```
1     for key,value in pairs(user_1)
2       do
3           print(key..value)
4       end
```

2. 函数

在 Lua 中，函数以 function 开头，以 end 结尾，funcName 是函数名，中间部分是函数体：

```
function funcName()
    ...
end
```

例如，自定义函数 contact()可以将两个字符串拼接：

```
1    function contact(str1,str2)
2        return str1..str2
3    end
```

5.4.3 应用案例

Redis 结合 Lua 脚本可以确保 Redis 命令执行时的顺序性和原子性，所以在开发一些高并发场景的应用时，可以采用两者结合的方式实现 Redis 事务，下面给出相关的范例。

1. 模拟抽奖

模拟一个抽奖场景，从奖池中进行随机抽奖。要求如下：

(1) 中奖的人只能从奖池中抽取。

(2) 每个人只能中奖一次。

(3) 中奖总人数不能超过设置的奖项数。

(4) 生成中奖名单。

可以使用 Redis 中的集合(SET)数据类型实现奖池。因为集合可以保证存放的元素无重复且无序，这样可以满足抽奖的随机性和奖池候选人的唯一性。同时集合还提供了很多操作来满足抽奖的需要。

第一步，设置奖池。向集合中添加参与抽奖的元素(人员)。代码如下：

```
1    public void preparation(){
2        for(int i = 1;i < 8;i++)
3            template.opsForSet().add("lottery","u" + i);
4    }
```

如上述代码所示，将集合 lottery 设置为奖池，并向奖池中添加 u1～u7 共 7 个元素。

第二步，编写抽奖脚本。从 lottery 集合中抽出一定数量的元素放入另外一个集合 chosen 中，遍历 chosen 集合即可得到中奖人员名单。这一功能需要执行多条 Redis 命令，可以通过编写 Lua 脚本实现。

可以在 Maven 项目的 src/main/resources 目录下创建名为 META-INF/scripts 的子目录，并将 lottery.lua 脚本文件存放在该子目录中，代码如文件 5-18 所示。

【文件 5-18】 lottery.lua

```
1     -- 奖池的 key
2    local lottery_key = KEYS[1]
3     -- 中奖名单的 key
4    local chosen_key = KEYS[2]
5     -- 预定抽奖的人数
6    local lottery_count = ARGV[1]
7
8     -- 如果预定抽奖的人数大于 0 则开始抽奖
9    if tonumber(lottery_count) > 0 then
10       -- 奖池中抽奖,返回的是被抽中的人组成的数组
11       local chosen_list = redis.call('SRANDMEMBER', lottery_key,
```

```
12            lottery_count);
13        -- 将抽中的人添加到中奖名单中,返回中奖的人数
14        if chosen_list then
15            return redis.call('SADD', chosen_key, unpack(chosen_list))
16        else
17            return 0
18        end
19    else
20        return 0
21    end
```

其中,KEYS 和 ARGV(第 2、4、6 行)是两个全局变量,它们可以将 Java 参数应用于 Lua 脚本。KEYS 表(数组)预先填充了在脚本执行之前提供给脚本的所有键名参数,而 ARGV 表(数组)的用途类似,但用于常规参数。

Lua 脚本与 Redis 交互可以通过调用 redis.call()函数或 redis.pcall()函数实现。这两个函数都可以执行格式正确的 Redis 命令(包括提供的参数)。它们之间的区别在于处理运行时错误(例如语法错误)的方式。执行 Redis 命令引发错误时,redis.call()直接返回给执行它的客户端。相反,redis.pcall()将返回到脚本的上下文,而不进行进一步处理。

例如,可以在 Redis 客户端利用 EVAL 命令执行 Lua 脚本:

```
redis > EVAL "return redis.call('SET',KEYS[1], ARGV[1])" 1 foo bar
OK
redis > get foo
"bar"
```

文件 5-18 的第 15 行的 unpack()函数接受一个数组作为参数,并返回数组的所有元素(数组下标默认从 1 开始)。在 Lua 5.1 中,unpack()是全局函数,可以直接使用,但是在 Lua 5.2 中,unpack()被移到 table.unpack,所以在 Lua 5.2 以后要用 table.unpack()替代 unpack()。

文件 5-18 的 Lua 脚本的返回值类型为 long。脚本的返回类型应为 long、boolean、list 或反序列化值类型之一,也可以为空。

第三步,调用抽奖脚本。Redis 2.6 及更高版本支持通过 EVAL 和 EVALSHA 命令运行 Lua 脚本。每当执行 EVAL 命令时,还会在请求中包含脚本的源代码。重复利用 EVAL 命令执行同一组参数化脚本时,既浪费网络带宽,又有一些 Redis 开销。为了节省网络和计算资源,Redis 为 Lua 脚本提供了缓存机制。

可以通过调用 RedisTemplate 的 execute()方法运行 Lua 脚本(该方法的全部重载形式可见表 3-7)。本案例选择的重载形式如下:

```
public < T > T execute(RedisScript < T > script, List < K > keys, Object... args) {
   return scriptExecutor.execute(script, keys, args);
}
```

其中:

(1) 参数 script 用来加载 Lua 脚本。

(2) 参数 keys 对应 Lua 脚本中的 KEYS,用来传入 Redis 的 KEY,在 Lua 脚本中可以通过 KEYS[索引]来取值,例如取第一个值 KEYS[1]。

(3) 参数 args 用来向 Lua 脚本传递其他的参数，在 Lua 脚本中可以通过 ARGV[索引]来取值。

调用 Lua 脚本的 Java 代码如下：

```
1    public void drawLotteries(RedisScript<Long> script, List<String> keys,
2        Object... args) {
3        template.execute(script, keys, args);
4    }
```

第四步，修改配置类。可修改文件 3-6，增加一个 RedisScript<Long> Bean 的定义，代码如下：

```
1    @Bean
2    public RedisScript<Long> script() {
3       RedisScript<Long> redisScript = RedisScript.of(new
4        ClassPathResource("META-INF/scripts/lottery.lua"), Long.class);
5       return redisScript;
6    }
```

第五步，编写测试代码。

```
1    @Test
2    public void testLottery() {
3       raffleDemo.preparation();
4       raffleDemo.drawLotteries(script, Arrays.asList("lottery","chosen")
5           , "3");
6       System.out.println(template.opsForSet().members("chosen"));
7    }
```

```
[u2, u6, u1]
```

图 5-13 抽奖案例运行结果

如测试代码所示，首先执行初始化工作，调用第一步定义的 preparation()方法设置奖池，随后从奖池 lottery 中随机抽取 3 个元素，并将这 3 个元素作为中奖者存入 chosen 集合。最后遍历 chosen 集合输出获奖者名单。运行测试代码后，控制台输出如图 5-13 所示。

2. 限流

限流是指某应用程序需要限制指定 IP 的主机在单位时间内的访问次数。互联网应用往往是高并发场景，其特性就是流量可能瞬时激增。此时，如果没有流量管控机制，就很容易导致系统崩溃。

限流是用来保证系统稳定性的常用手段，当系统遭遇瞬时流量激增时，很可能会因系统资源耗尽导致宕机。限流可以把超出系统承受能力的请求直接拒绝掉，允许其他请求正常访问，从而保证在系统可承受的范围内提供正常服务。例如，在某高并发场景中，对会员查询模块的限流需求是在 10 秒内最多允许有 1000 个请求。常用的限流算法有漏桶算法、令牌桶算法、计数法等。计数法限流的做法是，提供服务的模块会统计服务请求模块在单位时间内发送请求的次数，如果已达到限流标准，则不予提供服务。

下面给出用 Lua 脚本实现的基于计数法限流的案例。

第一步，创建执行限流的 Lua 脚本，代码如文件 5-19 所示。

【文件 5-19】 limiter.lua

```
1    local key = KEYS[1]
2    local limit = tonumber(KEYS[2])
3    local length = tonumber(KEYS[3])
4
5    local current = redis.call('GET', key)
6    if current == false then
7      redis.call('SET', key,1)
8      redis.call('EXPIRE',key,length)
9      return '1'
10   else
11     local num_current = tonumber(current)
12     if num_current + 1 > limit then
13        return '0'
14     else
15        redis.call('INCRBY',key,1)
16        return '1'
17     end
18   end
```

如文件 5-19 所示，该限流脚本有 3 个参数（第 1～3 行）：参数 key 接收待限流的对象，并将该对象的名字作为键保存到 Redis 中；参数 limit 表示限流的次数，即服务请求模块在单位时间内发送请求的次数如果不超过 limit，则其发出的请求可被正常处理；参数 length 表示限流的时长，即在 length 时间单位内只允许服务请求模块最多发出 limit 次请求。首先利用 Redis 的 GET 命令获取待限流的对象的访问次数（第 5 行）。如果获取不到（第 6～9 行），则认为当前对象第一次发送请求，在记录当前对象发送请求的次数后（第 7 行），设置当前请求的计数键在 Redis 中的过期时间（第 8 行），并返回字符串“1”（第 9 行），用以表示当前请求可以被正常处理。如果可以获取到待限流对象的访问次数（第 10～18 行），则首先调用 Lua 的 tonumber()函数将字符串形式的计数值转为数值型（第 11 行），随后判断当前计数值是否达到限流标准。如果达到限流标准（第 12、13 行），则返回字符串“0”，用以表示当前请求要被限流，不予提供服务；如果没达到限流标准（第 14～17 行），将当前计数值增 1（第 15 行），用以表示待限流对象在 length 时间内又一次发送请求，并返回字符串“1”。

第二步，以多线程的形式模拟多个并发请求，每个线程在规定时间内发送 5 次请求，以此查看限流效果，代码如文件 5-20 所示。

【文件 5-20】 LimiterDemo.java

```
1    public class LimiterDemo extends Thread {
2       private StringRedisTemplate template;
3       private RedisScript < String > script;
4       public LimiterDemo(StringRedisTemplate template,
5            RedisScript < String > script){
6          this.template = template;
7          this.script = script;
8       }
9
10      public void run(){
11         List < String > keyList = Arrays.asList(
```

```
12                Thread.currentThread().getName(),"3","10");
13        for(int i = 0;i < 5;i++){
14            String res = template.execute(script, keyList, "none");
15            if (res.equals("1"))
16                System.out.println(Thread.currentThread().getName() +
17                  " can visit");
18            else
19                System.out.println(Thread.currentThread().getName() +
20                  " sorry! be slow");
21        }
22    }
23 }
```

如文件 5-20 所示，以当前线程的名字作为键，将其发送请求的计数值存入 Redis。同时，指定每个线程在 10 秒内可以发送 3 次请求，如果在 10 秒的限流期内发送请求的次数超过 3 次，则该线程将被限流。Redis 缓存键（线程名）的时长设置为 10 秒（第 11、12 行）。而每个线程在程序运行期间将发送 5 个请求（第 13 行）。对照之前的分析可知，每个线程都会被执行两次限流。

第三步，为测试用多线程模拟多客户端发送请求，并被限流的效果，还需要编写测试代码，如文件 5-21 所示。

【文件 5-21】 TestLimiterDemo.java

```
1  @RunWith(SpringJUnit4ClassRunner.class)
2  @ContextConfiguration(classes = RedisConfig.class)
3  public class TestLimiterDemo {
4      @Autowired
5      private StringRedisTemplate template;
6      @Autowired
7      private RedisScript < String > script;
8
9      @Test
10     public void testLimiter() {
11         try {
12             for(int i = 0;i < 5;i++) {
13                 Thread.sleep(200);
14                 new LimiterDemo(template, script).start();
15             }
16            //主线程挂起，等待所有线程运行完毕
17            Thread.sleep(6000);
18         } catch (InterruptedException e) {
19                 e.printStackTrace();
20             }
21     }
22 }
```

如文件 5-21 所示，第 2 行引入配置类 RedisConfig，增加一个 RedisScript < String > Bean 的定义，代码可参考模拟抽奖案例的第四步，此处不再赘述。随后创建 5 个线程（第 12～15 行），模拟 5 个客户端给服务模块发送请求，测试限流效果，代码运行结果如图 5-14 所示。

```
Thread-1 can visit
Thread-0 can visit
Thread-1 can visit
Thread-0 can visit
Thread-1 can visit
Thread-0 can visit
Thread-1 sorry! be slow
Thread-0 sorry! be slow
Thread-1 sorry! be slow
Thread-0 sorry! be slow
Thread-2 can visit
Thread-2 can visit
Thread-2 can visit
Thread-2 sorry! be slow
Thread-2 sorry! be slow
Thread-3 can visit
Thread-3 can visit
Thread-3 can visit
Thread-3 sorry! be slow
Thread-3 sorry! be slow
Thread-4 can visit
Thread-4 can visit
Thread-4 can visit
```

图 5-14 限流案例代码运行结果

5.5 Geo

Redis 3.2 及以后的版本提供了 Geo(Geospatial,地理空间数据)类型,该类型可以存储地理位置信息,用来实现诸如附近位置、摇一摇等这类依赖于地理位置信息的功能。Redis Geo 并不是一种新的数据结构,而是基于有序集合实现的。有序集合保存的数据形式是键-分数(Key-Score),即一个元素对应一个分数,默认是根据分数排序的,且可以按范围查询。但是经纬度是一个数据对,如(106.053 218,32.437 117),Redis 是如何将经纬度转换为分数值的呢?转换为分数值之后,又是如何保证分数值相邻的元素距离也相近的呢?这一切依赖于 GeoHash 编码。该编码可将经纬度转换为字符串,而且距离越近字符串的相似程度越高。GeoHash 编码可通过 Redis 提供的 Geo 相关命令得到。Geo 相关命令包括添加、计算位置之间的距离,根据中心点坐标和距离范围来查询地理位置集合等,如表 5-6 所示。

表 5-6 Geo 相关命令

命 令	说 明
GEOADD	添加地理位置的坐标
GEOPOS	获取地理位置的坐标
GEODIST	计算两个位置之间的距离
GEORADIUS	根据用户给定的经纬度坐标来获取指定范围内的地理位置集合
GEORADIUSBYMEMBER	根据存储在位置集合里面的某个地点获取指定范围内的地理位置集合
GEOHASH	返回有效的 GeoHash 字符串,该字符串表示一个或多个元素在表示地理空间索引的排序集中的位置(其中元素是使用 GEOADD 添加的)

操作 Redis Geo 的方法与操作 Redis 字符串的方法类似,要调用 RedisTemplate 的 opsForGeo()方法创建 GeoOperations<K,M>子接口对象,再调用 GeoOperations<K,M>子接口的相关方法。下面介绍 GeoOperations<K,M>子接口中的常用方法。

（1）方法原型：@Nullable Long add(K key,Point point,M member)；功能：将具有给定成员名称的点(地理坐标值)添加到键 key；对应的 Redis 命令：GEOADD。其中，Point类(org. springframework. data. geo. Point)代表某点的地理坐标值；泛型 M 为 GeoOperations 接口的类型参数。

（2）方法原型：@Nullable Distance distance(K key,M member1,M member2,Metric metric)；功能：获取成员 member1 和成员 member2 之间的距离；对应的 Redis 命令：GEODIST。其中，Metric 接口(org. springframework. data. geo. Metric)是一个用于米制的度量接口。

（3）方法原型：@Nullable List < Point > position(K key,M...members)；功能：获取一个或多个成员的位置的点(地理坐标值)表示；对应的 Redis 命令：GEOPOS。

（4）方法原型：@Nullable GeoResults < RedisGeoCommands. GeoLocation < M >> radius(K key,M member,Distance distance)；功能：获取给定圆边界内的成员；对应的 Redis 命令：GEORADIUS。

（5）方法原型：@Nullable GeoResults < RedisGeoCommands. GeoLocation < M >> radius(K key,M member,Distance distance)；功能：基于米制单位，获取由成员坐标和给定半径定义的圆内的成员；对应的 Redis 命令：GEORADIUSBYMEMBER。

（6）方法原型：@Nullable List < String > hash(K key,M...members)；功能：获取一个或多个成员的位置的 GeoHash 表示；对应的 Redis 命令：GEOHASH。

案例：结合表 5-7 给出的城市的地理位置信息，完成如下要求。

表 5-7　5 个城市的地理位置信息

城　　市	经　　度	纬　　度	成　员　名
北京	116.28	39.55	BeiJing
天津	117.12	39.08	TianJin
石家庄	114.29	38.02	ShiJiaZhuang
唐山	118.01	39.38	TangShan
保定	115.29	38.51	BaoDing

（1）以 city:location 为键，将表 5-7 中的城市位置信息加入 Redis。可以调用 GeoOperations <K,M>子接口的 add()方法，测试代码如文件 5-22 所示。

【文件 5-22】　TestGeoDemo. java

```
1    @RunWith(SpringJUnit4ClassRunner.class)
2    @ContextConfiguration(classes = RedisConfig.class)
3    public class TestGeoDemo {
4       @Autowired
5       private RedisTemplate template;
6       private double[][] position = {{116.28,39.55},{117.12,39.08},
7             {114.29,38.02},{118.01,39.38},{115.29,38.51}};
8       private String[] member =
9         {"BeiJing","TianJin","ShiJiaZhuang","TangShan","BaoDing"};
10
11      @Test
12      public void testGeoAdd(){
```

```
13            Long num = 0L;
14            for(int i = 0;i < member.length; i++){
15                Point point = new Point(position[i][0],position[i][1]);
16                num = template.opsForGeo()
17                    .add("city:location",point,member[i]);
18            }
19            Assert.assertEquals(1,num.intValue());
20        }
21    }
```

(2) 获取天津的位置信息并输出。可调用 GeoOperations <K,M>子接口的 position() 方法,在文件 5-22 中添加一个测试用例,代码如下。

```
1    @Test
2    public void testGeoPos(){
3        List<Point> list = new ArrayList<Point>();
4        list = template.opsForGeo().position("city:location",member[1]);
5        list.forEach(System.out::println);
6    }
```

该测试代码的运行结果如图 5-15 所示。

```
Point [x=117.120000, y=39.080000]
```

图 5-15 (2)测试代码运行结果

(3) 计算北京和天津间的直线距离,单位: km。可调用 GeoOperations <K,M>子接口的 distance()方法,并设置计算结果的单位为千米。在文件 5-22 中添加一个测试用例,代码如下。

```
1    @Test
2    public void testGeoDist() {
3        Distance distance = template.opsForGeo().distance("city:location",
4            member[0],member[1],RedisGeoCommands.DistanceUnit.KILOMETERS);
5        System.out.println(distance);
6    }
```

其中,第 4 行的参数 RedisGeoCommands.DistanceUnit 是一个静态枚举,它已实现 Metric 接口。从 RedisGeoCommands.DistanceUnit 选择常量 KILOMETERS 用来指定计算结果的单位为 km。完成要求(3)也可以使用 distance()方法的另一种重载形式,即省略第 4 行中的第 3 个参数,这样计算出的两个成员间的距离单位为 m。

(4) 找出距离北京 150km 以内的城市。可调用 GeoOperations <K,M>子接口提供的 radius()方法。在文件 5-22 中添加一个测试用例,代码如下。

```
1    @Test
2    public void testGeoRadius() {
3        Distance distance = new Distance(150, Metrics.KILOMETERS);
4        GeoResults<RedisGeoCommands.GeoLocation<String>> results =
5            template.opsForGeo().radius("city:location",member[0],distance);
6        System.out.println(results);
7    }
```

此测试代码的运行结果如图 5-16 所示。

当然,也可使用 radius()方法的另一种重载形式: @Nullable。

```
GeoResults: [averageDistance: 0.0 KILOMETERS, results: GeoResult
 [content: RedisGeoCommands.GeoLocation(name=BeiJing, point=null),
 distance: 0.0 KILOMETERS, ],GeoResult [content: RedisGeoCommands
 .GeoLocation(name=TianJin, point=null), distance: 0.0 KILOMETERS,
 ],GeoResult [content: RedisGeoCommands.GeoLocation(name=TangShan,
 point=null), distance: 0.0 KILOMETERS, ],GeoResult [content:
 RedisGeoCommands.GeoLocation(name=BaoDing, point=null), distance:
 0.0 KILOMETERS, ]]
```

图 5-16 （4）测试代码运行结果

GeoResults < RedisGeoCommands. GeoLocation < M >> radius(K key,Circle within)完成此要求。读者可自行编写代码。

（5）利用键绑定子接口操作键为 bound-geo 的 Geo 类型的数据。分别执行添加、测距操作。例如，测量 BeiJing 和 BaoDing 两个城市间的距离，单位为 km，部分代码如下。

```
1    @Test
2    public void testGeoBoundOperations() {
3        BoundGeoOperations < String, String > ops =
4            template.boundGeoOps("bound - geo");
5        ops.add(new Point(116.28,39.55),"BeiJing");
6        ops.add(new Point(115.29,38.51),"BaoDing");
7        Distance dis = ops.distance("BeiJing","BaoDing",
8            RedisGeoCommands.DistanceUnit.KILOMETERS);
9        if(dis != null)
10           System.out.println("The distance between BeiJing and BaoDing
11           is " + dis.getValue() + dis.getUnit());
12   }
```

此测试代码的运行结果如图 5-17 所示。

```
The distance between BeiJing and BaoDing is 143.8646km
```

图 5-17 （5）测试代码运行结果

（6）获取北京和天津的 GeoHash 值。可调用 GeoOperations < K,M >子接口的 hash()方法。在文件 5-22 中添加一个测试用例，代码如下。

```
1    @Test
2    public void testGeoHash() {
3        List < String > list = template.opsForGeo()
4            .hash("city:location",member[0],member[1]);
5        list.forEach(System.out::println);
6    }
```

```
wx48ypbe2q0
wwgq34k1tb0
```

图 5-18 （6）测试代码运行结果

Redis 使用 GeoHash 将二维经纬度转换为一维字符串。上述测试代码的运行结果如图 5-18 所示。

由上述案例可知，GeoHash 具有如下特点：

（1）Geo 的基础数据类型为有序集合，Redis 将所有地理位置信息的 GeoHash 存放在有序集合中。

（2）字符串越长，表示的位置越精确。例如，GeoHash 的长度为 9 时，精度在 2m 左右。

（3）两个字符串越相似，它们之间的距离越近。Redis 利用字符串前缀匹配算法实现相关的命令。

(4) GeoHash 编码和经纬度是可以相互转换的。

5.6 小结

本章介绍了 Redis 的一些基础(非缓存)应用。如 Redis 可以实现消息的发布-订阅,可以作为简单的消息队列使用。此外,Redis 还提供了以日志形式存储消息的数据结构——流。这样,就支持消费者以同步或异步方式获取消息队列中的消息。为了提高命令的执行效率,Redis 提供了流水线机制。它允许将多条命令集中起来,一次性发给 Redis 服务器执行,而后一次性返回命令执行结果。作为 NoSQL 数据库,Redis 只提供了简单的事务管理功能,结合了 Lua 脚本后,Redis 才拥有强大的事务管理能力。最后,本章介绍了 Redis 提供的基于有序集合的新的数据类型——Geo。利用 Geo 类型,可以执行地理位置相关的计算。

对于以上 5 方面的基础应用,Redis 都提供了相关命令。读者可查找 Redis 官方提供的命令手册学习更多相关命令的使用方法。

第 6 章

视频讲解

响应式Redis

响应式编程(Reactive Programming,RP)是一种面向数据流和传播变化的新的编程范式,是为了实现复杂的事件响应系统而提出的一种编程思想,目的是简化复杂事件响应系统的开发。借助响应式编程,可以在编程语言中很方便地表达静态或动态的数据流,而相关的计算模型会自动将变化的值通过数据流进行传播。与响应式编程相对应的是命令式编程。目前大多数的程序设计语言采用的是命令式编程。在命令式编程中,当调用方法 m()进行数据库读写操作时,这类方法不能立即返回,也无法中断,需要等待数据库读写操作完成才会返回,因此一个线程只能处理一个请求。在 m()方法执行过程中,CPU 需要等待操作结果返回以进行后续处理,这样 CPU 资源就没有很好地利用起来。而在响应式编程中,调用方法 m()后无须等待操作结果,可以立刻返回。这样 CPU 可以继续执行后续指令。m()方法的执行结果可以以流的方式返回。一旦 m()方法的操作结果就绪,处理即可。这样就极大提高了 CPU 的利用率。

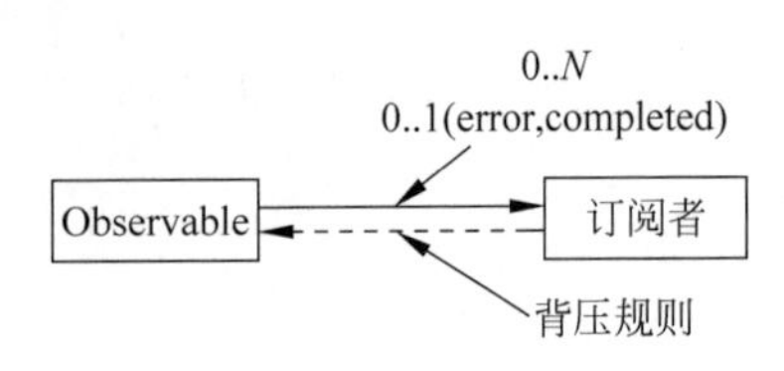

图 6-1　响应式编程的基础模型

响应式编程本质上是对观察者模式的高度封装。图 6-1 展示了响应式编程的基础模型。其中,Observable 定义了发布者(Publisher),发布者负责在其状态改变时发出通知。通知的形式可以是多项数据(0～N 项数据,以 0..N 表示),也可以是单项数据(0 项或 1 项数据,以 0..1 表示)。Subscriber 是订阅者,订阅者负责观察发布者的状态变化并做出响应。订阅者必须订阅一个发布者才能接收通知,而且订阅者只具备简单的数据处理接口和异常处理接口。背压规则(Backpressure Regulation)的作用是当订阅者的处理速度跟不上发布者时,背压规则向发布者发出反馈信号,控制发布者放慢消息发送速度。这样可以避免订阅者因处理速度的差异直接丢弃数据,从而导致大量数据丢失。

目前,比较成熟的响应式编程库是 Reactor。它是一个基于 JVM 之上的异步应用基础库。Reactor 的基本设计思想就是当执行一个带有一定延迟才能够返回的操作时,程序不会阻塞,而是立刻返回一个流,并且订阅这个流。当这个流上产生了返回数据,可以立刻得到通知并调用回调函数处理返回的数据。就是说,把程序同步执行方式换成异步,方法执行变成消息发送。这就是响应式编程的重要特性。此外,响应式编程还有以下特点。

(1) 异步编程。提供了合适的异步编程模型,能够挖掘多核 CPU 的能力,提高效率、降低延迟和阻塞。

(2) 数据流。基于数据流模型,响应式编程提供一套统一的流风格的数据处理接口。

除了支持静态数据流外，还支持动态数据流，并且允许复用和同时接入多个订阅者。

(3) 传播变化。简单来说就是以一个数据流为输入，经过一系列转换后得到另一个数据流，然后分发给各个订阅者。这个过程类似于函数式编程中的组合函数，将多个函数串联起来，把一组输入数据转换为格式迥异的输出数据。

6.1 Reactor 简介

6.1.1 Reactor 库

Reactor 库是一个基于 JVM 之上的非阻塞响应式编程应用基础库，实现了响应式流规范。它为 Java、Groovy 和其他 JVM 语言提供了构建基于事件和数据驱动应用程序的抽象库。Reactor 已与 Java 8 API 集成，特别是 CompletableFuturel 类、Stream 接口和 Duration 类。要使用 Reactor 的核心功能，Java 的最低版本为 Java 8。Reactor 有很好的性能，在最新的硬件平台上，使用无阻塞分发器每秒可处理 1500 万个事件。

Reactor 提供了实现发布者的响应式类 Flux(reactor. core. publisher. Flux < T >)和 Mono(reactor. core. publisher. Mono < T >)，以及丰富的操作符。Flux 代表 0～*N* 个元素的响应式流，Mono 代表 0 或 1 个元素的响应式流。

Flux 和 Mono 之间可以转换，如：Flux 的 count 操作(计算流中元素个数)返回 Mono，Mono 的 concatWith 操作(连接另一个响应式流)返回 Flux。

6.1.2 Publisher

由于响应式流的特点，程序不能再返回一个简单的 POJO(Plain Ordinary Java Object，普通的 Java 对象)来表示结果了，必须返回一个类似 Java 中的 Future(java. util. concurrent. Future)的对象，在有结果可用时通知消费者消费响应数据。在响应式流(Reactive Stream)规范中，这种返回结果被定义为 Publisher < T >。Publisher < T >是一个可以提供 0～*N* 个元素序列的发布者，并根据其订阅者 Subscriber <? super T >的需求推送元素。一个发布者可以支持多个订阅者，并可以根据订阅者的逻辑推送元素。

6.1.3 Flux

Flux < T >是一个能够发出 0～*N* 个元素的异步标准发布者。发布消息的过程会被一个完成(completion)或错误(error)信号终止。因此，一个 Flux < T >可能发布的结果是响应数据、completion 和 error，这三个结果会传递给订阅者的 onNext()、onComplete()和 onError()方法分别进行处理。所有的信号事件，包括终止发布的信号事件都是可选的。如果没有 onNext 事件，但是有 onComplete 事件，那么 Flux 发出的就是空的有限流；如果去掉 onComplete 事件就得到一个无限的空数据流。无限的数据流也可以不是空的，如 Flux. interval(Duration)生成的是一个 Flux < Long >，这是一个无限周期性发出规律整数的时钟数据流。

6.1.4 Mono

Mono < T >是一种特殊的发布者，它每次最多发布一个元素，然后(可选的)终止于

onComplete（完成）或 onError（错误）信号。

Mono 中的操作符是 Flux 中操作符的子集，即 Flux 中只有部分操作符适用于 Mono，有些操作符是将 Mono 和另一个发布者连接转换为 Flux。例如，Mono # concatWith（Publisher）转换为 Flux，Mono # then（Mono）返回另一个 Mono。可以使用 Mono < Void > 来创建一个只有完成概念的空值异步处理过程（类似于 Runnable），即 Mono < Void >只发布一个代表操作完成的空数据流。

在 Spring Data Redis 中，使用的响应库是 Reactor。本章后续内容将介绍怎样利用 Flux 和 Mono 来实现响应式编程操作 Redis。

6.2 响应式 Redis 基础

Spring Data Redis 同时支持 Jedis 客户端和 Lettuce 客户端，但是仅 Lettuce 客户端才支持响应式操作。Spring Data Redis 提供的与响应式相关的主要接口或类有：

（1）ReactiveRedisConnection 接口（org. springframework. data. redis. connection. ReactiveRedisConnection）：该接口是与 Redis 通信的核心，它采用响应式方式连接 Redis，并提供响应式命令的执行入口。

（2）ReactiveRedisConnectionFactory 接口（org. springframework. data. redis. connection. ReactiveRedisConnectionFactory）：该接口用来创建 ReactiveRedisConnection 实例。此外，此接口还继承了 PersistenceExceptionTranslator 接口，这意味着工厂一旦声明，就可以进行透明的异常转换。例如，通过使用@Repository 注解和 AOP 进行异常转换。

（3）ReactiveRedisOperations 接口（org. springframework. data. redis. core. ReactiveRedis-Operations < K，V >）：该接口指定了一组基本的 Redis 操作，这些操作由该接口的实现类 ReactiveRedisTemplate 实现。

（4）ReactiveRedisTemplate 类（org. springframework. data. redis. core. ReactiveRedis Temp late< K，V >）：该类是实际的响应式操作类，与 RedisTemplate 一样，提供了 opsFor *** ()方法，使用这些方法可以获取字符串、哈希、集合、有序集合、列表、地理空间数据等类型的响应式操作接口，如表 6-1 所示。

表 6-1 用于操作 Redis 特定类型的响应式接口

接　口	描　述
ReactiveGeoOperations	执行地理空间操作，如 GEOADD、GEORADIUS 等
ReactiveHashOperations	执行哈希操作
ReactiveHyperLogLogOperations	执行 HyperLogLog 操作，如 PFADD、PFCOUNT 等
ReactiveListOperations	执行列表操作
ReactiveSetOperations	执行集合操作
ReactiveValueOperations	执行字符串（值）操作
ReactiveZSetOperations	执行有序集合操作

下面介绍一个简单的响应式应用程序案例，以响应方式实现 Redis 读写。

第一步，编写配置类，它配置了一个 ReactiveRedisTemplate 响应式操作类，代码如文件 6-1 所示。

【文件 6-1】 MyFirstReactiveConfiguration.java

```
1    @Configuration
2    public class MyFirstReactiveConfiguration {
3    @Bean
4    public ReactiveRedisConnectionFactory lettuceConnectionFactory() {
5      LettuceClientConfiguration clientConfig =
6      LettuceClientConfiguration.builder()
7       .commandTimeout(Duration.ofSeconds(2))
8       .shutdownTimeout(Duration.ZERO).build();
9       return new LettuceConnectionFactory(new
10        RedisStandaloneConfiguration("localhost", 6379), clientConfig);
11   }
12   @Bean
13   public ReactiveRedisTemplate<String, String> reactiveRedisTemplate(
14        ReactiveRedisConnectionFactory factory) {
15      RedisSerializationContext
16      .RedisSerializationContextBuilder<String, String>
17        builder = RedisSerializationContext
18         .newSerializationContext();
19      //设置序列化方式
20      builder.key(new StringRedisSerializer());
21      builder.value(RedisSerializer.string());
22      builder.hashKey(new StringRedisSerializer());
23      builder.hashValue(new StringRedisSerializer());
24      builder.string(new StringRedisSerializer());
25      RedisSerializationContext<String,String> build = builder.build();
26      return new ReactiveRedisTemplate<>(factory, build);
27     }
28   }
```

如文件 6-1 所示，第 4～11 行使用 Lettuce 客户端创建与 Redis 的连接。在创建 Redis 连接的同时，可以设置 Lettuce 客户端的连接属性。如第 7 行设置命令的超时时长，第 8 行设置关机超时时长。ReactiveRedisTemplate 的大部分操作都使用基于 Java 的序列化器。ReactiveRedisTemplate 写入或读取的任何对象都通过 RedisElementWriter 或 RedisElementReader 进行序列化或反序列化。第 20～24 行设置序列化方式，StringRedisSerializer 支持 String 类型数据和 byte[]数据之间的序列化和反序列化。

第二步，获取操作 Redis 的字符串类型数据的响应式接口，然后利用该接口设置和获取指定键的值，代码如文件 6-2 所示。

【文件 6-2】 MyReactiveRedisService.java

```
1    @Service
2    public class MyReactiveRedisService {
3      @Autowired
4      private ReactiveRedisTemplate<String, String> redisTemplate;
5      public Mono<Boolean> set(String key, String value, Long time) {
6         return redisTemplate.opsForValue().set(key, value, Duration.
7          ofSeconds(time));
8      }
9      public Mono<String> get(String key) {
```

```
10          return key == null ? null : redisTemplate.opsForValue().get(key);
11      }
12  }
```

如文件 6-2 所示，第 5～8 行定义了向 Redis 中写入数据的操作。其中第 6 行调用 ReactiveValueOperations < K,V >接口的 set()方法，设置了键 key 的值 value 和键过期超时时长 time。第 9～11 行定义了从 Redis 中获取指定键对应的值的操作。其中第 10 行调用了 ReactiveValueOperations < K,V >接口的 get()方法，与第 6 行调用的 set()方法类似，这两个方法都在执行 Redis 的读写操作。为避免读写操作造成程序阻塞，这两个方法的返回值都为 reactor.core.publisher.Mono < T >，即创建一个消息发布者，允许程序无须等待读写操作结果，以异步方式继续执行，直到收到读写操作完成的信号再进行后续处理。

第三步，编写测试代码，订阅上述代码中 Mono 发布的消息，部分测试代码如文件 6-3 所示。

【文件 6-3】 TestMyReactiveRedisService.java

```
1   @RunWith(SpringJUnit4ClassRunner.class)
2   @ContextConfiguration(classes = MyFirstReactiveConfiguration.class)
3   public class TestMyReactiveRedisService {
4     @Resource
5     private MyReactiveRedisService redisService;
6
7     @Test
8     public void testMyReactiveRedisOperations() {
9       //保存 5 分钟
10      redisService.set("test", "test_value", 5 * 60L)
11        .subscribe(System.out::println);
12      redisService.get("test").subscribe(System.out::println);
13    }
14  }
```

如文件 6-3 所示，第 10 行首先以 test 为键，将值 test_value 写入 Redis，并指定该键值在 Redis 中保存 5 分钟。由于执行了 Redis 的读写操作，程序可能会因此而阻塞。第 12 行调用 subscribe()方法订阅读写操作的返回消息。当读写操作执行结束后，用 Lambda 表达式向控制台输出读写操作的返回消息。测试代码运行后的结果如图 6-2 所示。

```
true
test_value
```

图 6-2 文件 6-3 测试代码运行结果

6.3 使用 ReactiveStringRedisTemplate

由于存储在 Redis 中的键和值通常是 java.lang.String 类型，可以将 ReactiveRedisTemplate 的泛型设置为< String, String >（如文件 6-1 的第 13 行）。此外，Spring Data Redis 为 ReactiveRedisTemplate 类提供了基于字符串的扩展：ReactiveStringRedisTemplate 类。可以使用这个类完成密集的字符串操作。除了绑定到字符串类型的键以外，该类还使用了基于字符串的序列化上下文（RedisSerializationContext，如文件 6-1 的第 25 行），这意味着 Redis 中存储的键和值是人类可以读懂的（假设 Redis 和应用程序代码使用了相同的编码规则)。文件 6-4 展示了如何配置 ReactiveStringRedisTemplate。

【文件 6-4】 MyReacConf.java

```
1   @Configuration
2   public class MyReacConf {
3       @Bean
4       public ReactiveRedisConnectionFactory connectionFactory() {
5           return new LettuceConnectionFactory("localhost", 6379);
6       }
7       @Bean
8       public ReactiveStringRedisTemplate reactiveStringRedisTemplate(
9        ReactiveRedisConnectionFactory factory) {
10           return new ReactiveStringRedisTemplate(factory);
11      }
12  }
```

本节介绍如何使用 ReactiveStringRedisTemplate 类响应式操作 Redis 中的字符串、列表、哈希、集合、有序集合和地理空间数据类型的数据。本节约定,键和值的类型均为 String。操作上述类型的响应式接口见表 6-1。

6.3.1 操作字符串类型的数据

Spring Data Redis 中,操作字符串类型数据的响应式接口为 ReactiveValueOperations<K,V>,该接口定义的方法和 ValueOperations、BoundValueOperations 接口定义的方法类似。可以通过 ReactiveRedisTemplate(或 ReactiveStringRedisTemplate)类的 opsForValue()方法获取 ReactiveValueOperations 对象,代码如下:

```
ReactiveValueOperations<String, String> ops = redisTemplate.opsForValue();
```

其中,ReactiveValueOperations 接口的主要方法如下。

1. 设置值和获取值

用于在 Redis 中设置和获取值的响应式方法有:

(1) reactor.core.publisher.Mono<Boolean> set(K key,V value):将给定的值 value 设置到键 key。

(2) reactor.core.publisher.Mono<Boolean> set(K key,V value,Duration timeout):将给定的值 value 设置到键 key,且指定过期时间 timeout。

(3) reactor.core.publisher.Mono<V> get(Object key):获取指定键 key 的值。

(4) reactor.core.publisher.Mono<V> getAndSet(K key,V value):将值 value 设置到键 key,且返回旧的值。

(5) reactor.core.publisher.Mono<List<V>> multiGet(Collection<K> keys):批量获取键集合 keys 对应的所有值。

(6) reactor.core.publisher.Mono<Boolean> multiSet(Map<? extends K,? extends V> map):将参数 map 中所有的键值对写入 Redis。

【例 6-1】 以响应式操作将值 value0 以键 value-key0 保存到 Redis,代码如文件 6-5 所示(注:6.3 节例题如无特殊说明,均指执行响应式操作)。

【文件 6-5】 TestReactiveValueOperations. java

```
1    @RunWith(SpringJUnit4ClassRunner.class)
2    @ContextConfiguration(classes = MyReacConf.class)
3    public class TestReactiveValueOperations {
4      @Autowired
5      private ReactiveStringRedisTemplate tmp;
6      @Test
7      public void testReactiveValueOperations(){
8        ReactiveValueOperations<String,String> ops = tmp.opsForValue();
9        ops.set("value-key0","value0").subscribe();
10       ops.get("value-key0").subscribe(System.out::println);
11     }
12   }
```

如文件 6-5 所示，第 8 行通过调用 ReactiveRedisTemplate 类的 opsForValue()方法获取 ReactiveValueOperations 对象。向 Redis 中存入字符串数据，可调用 ReactiveValueOperations 接口的 set()方法(第 9 行)，该方法会返回一个 reactor. core. publisher. Mono < Boolean >对象。作为发布者对象，Mono 会对外发布 set()方法执行成功与否的消息。因此，程序需要向此 Mono 对象订阅 set()方法执行的结果以便进行后续处理。第 9 行调用了 Mono 类的 subscribe()方法实现订阅。Mono 对象的时间线如图 6-3 所示。当调用 subscribe()方法时，相当于对 Mono 对象执行一次转换。如果 Mono 有消息可供发布，则返回该消息，并成功完成。如果 Mono 对象没有要发布的消息，则 Mono 对象不会正常终止。在执行 set()方法向 Redis 写入字符串数据后，可以通过指定键将写入的值取出，如第 10 行调用 get()方法。此处调用的是 subscribe()方法的另一个重载形式，它的原型如下：

```
public final Disposable subscribe(Consumer<? super T> consumer)
```

可见，该方法的参数为一个消费者接口。因此，可以使用方法引用将 get()方法的执行结果输出到控制台。

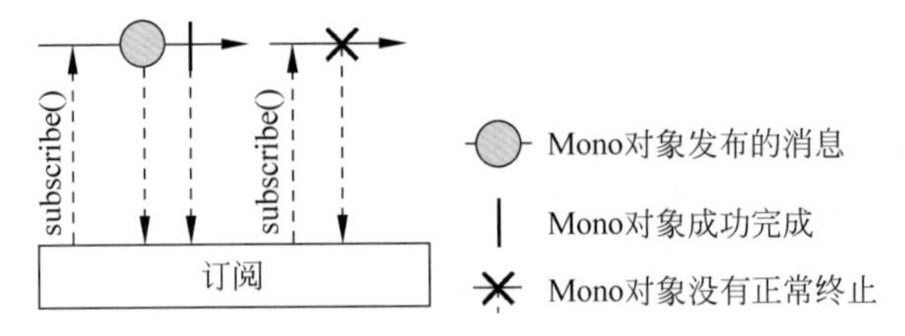

图 6-3 Mono 对象的时间线

【例 6-2】 将值 value1、value2 和 value3 一次性保存到 Redis 中，键分别为 value-key1、value-key2 和 value-key3。再利用这 3 个键一次性将对应的值取出并输出到控制台。可在文件 6-5 基础上增加测试代码，如文件 6-6 所示。

【文件 6-6】 例 6-2 测试代码

```
1    @Test
2    public void testReactiveValueOperations(){
3      Map<String,String> map = new HashMap<>();
4      map.put("value-key1", "value1");
5      map.put("value-key2", "value2");
```

```
6        map.put("value-key3", "value3");
7        ops.multiSet(map).subscribe();
8        ops.multiGet(Arrays
9           .asList("value-key3","value-key2","value-key1"))
10          .subscribe(list -> list.forEach(System.out::println));
11     }
```

如文件 6-6 所示，在第 8 行执行了 multiGet()方法后，会返回一个 reactor. core. publisher. Mono < java. util. List < V >>与对象，与例 6-1 类似，此处可调用 Mono 对象的带参数的 subscribe()方法完成对操作结果消息的订阅。这个方法的参数是一个消费者接口。在获取到操作结果后，用 Lambda 表达式向控制台输出结果。

2. 追加值

用于在 Redis 中追加值的响应式方法有：

reactor. core. publisher. Mono < Long > append(K key, String value)：将给定的值 value 追加到键 key 的值的尾部。

【例 6-3】 设置键 append-key 的值为 hello，操作结束后向该键追加值 redis。在控制台输出该键的新值。测试代码如文件 6-7 所示。

【文件 6-7】 例 6-3 测试代码

```
1     @Test
2     public void testReactiveValueOperations(){
3        ops.set("append-key","hello").subscribe();
4        ops.append("append-key"," redis").subscribe();
5        System.out.println("新值: ");
6        ops.get("append-key").subscribe(System.out::println);
7     }
```

如文件 6-7 所示，向既有键追加值，可调用 ReactiveValueOperations 接口的 append() 方法。与例 6-1、例 6-2 类似，在以响应式执行 Redis 的读写操作后，要调用 subscribe()方法（包括其重载形式）订阅操作结果，以便进行后续处理。

3. 取字符串长度和删除

用于获取 Redis 中字符串长度和删除指定键值的响应式方法有：

(1) reactor. core. publisher. Mono < Long > size(K key)：获取指定键 key 存储值的长度。

(2) reactor. core. publisher. Mono < Boolean > delete(K key)：删除给定的键 key。

【例 6-4】 设置键 obtain-key 的值为 hello redis，操作结束后向控制台输出该键的长度并删除该键。测试代码如文件 6-8 所示。

【文件 6-8】 例 6-4 测试代码

```
1     @Test
2     public void testReactiveValueOperations(){
3       ops.set("obtain-size","hello,redis").subscribe();
4       System.out.print("The size of hello redis is ");
5       ops.size("obtain-size").subscribe(System.out::println);
6       //delete
```

```
7       ops.delete("obtain-size").subscribe(aBoolean -> System.out.println
8         (aBoolean?"删除成功":"删除失败"));
9   }
```

从上面的例子可知，要实现对消息发布者 Flux 或 Mono 的订阅，需要调用 Flux 或 Mono 类的 subscribe()方法，本节列举两个重载的 subscribe()方法：

（1）subscribe()：该方法用于订阅消息发布者发布的消息，并触发消息序列。

（2）subscribe(Consumer <? super T > consumer)：该方法除完成常规的订阅操作外，还对产生的每个消息值做相应的处理。该方法以消费者接口作为参数，因此可以使用 Lambda 表达式处理 subscribe()方法的参数。

subscribe()方法的所有这些变体返回对订阅的引用，当不需要更多数据时，可以使用该引用来取消订阅。取消后，源(发布者)应停止产生值，并清理其创建的任何资源。这种取消和清理行为在 Reactor 中由通用的 Disposable(reactor.core.Disposable)接口表示。可调用 Disposable 接口的 dispose()方法取消订阅。对于 Flux 或 Mono，取消订阅是指源(发布者)停止产生值的信号。然而，取消订阅不能保证立即生效：因为某些源可能会快速生成值，甚至在收到取消指令之前就已经完成。

6.3.2 操作列表类型的数据

Spring Data Redis 中，操作列表类型的响应式接口为 ReactiveListOperations < K,V >，该接口定义的方法和 ListOperations、BoundListOperations 接口定义的方法类似。

可以通过 ReactiveRedisTemplate(或 ReactiveStringRedisTemplate)类的 opsForList()方法获取 ReactiveListOperations 对象，代码如下：

```
ReactiveListOperations < String,String > ops = redisTemplate.opsForList();
```

其中，ReactiveListOperations 接口的主要方法如下。

1. 列表元素的添加与删除

用于在 Redis 列表中添加、删除元素的响应式方法有：

（1）reactor.core.publisher.Mono < Long > leftPush(K key,V value)：将值 value 添加到列表 key 的左边。

（2）reactor.core.publisher.Mono < Long > leftPushAll(K key,Collection < V > values)：将多个值 values 添加到列表 key 的左边。

（3）reactor.core.publisher.Mono < Long > rightPush(K key,V value)：将值 value 添加到列表 key 的右边。

（4）reactor.core.publisher.Mono < Long > rightPushAll(K key,Collection < V > values)：将多个值 values 添加到列表 key 的右边。

（5）reactor.core.publisher.Mono < V > leftPop(K key)：移除并返回列表 key 中的第一个元素。

（6）reactor.core.publisher.Mono < V > leftPop(K key,Duration timeout)：在指定的超时时间 timeout 内移除并返回列表 key 中的第一个元素。

（7）reactor.core.publisher.Mono < V > rightPop(K key)：移除并返回列表 key 的最

后一个元素。

(8) reactor. core. publisher. Mono < V > rightPop(K key,Duration timeout):指定在超时时间 timeout 内移除并返回列表 key 的最后一个元素。

(9) reactor. core. publisher. Mono < Long > remove(K key,long count,Object value):从列表中移除首次出现的 count 个值为 value 的元素。

【例 6-5】 将值 value1 添加到列表 list-key1 的左边并显示操作结果。随后,将值 value2、value3 和 value4 批量添加到 list-key1 的左边并显示操作结果。随后,移除并返回存储在 list-key1 中的第一个元素。测试代码如文件 6-9 所示。

【文件 6-9】 例 6-5 测试代码

```
1    @Test
2    public void testLeftPush() throws InterruptedException {
3        ops.leftPush("list - key1","value1").subscribe(
4            aLong - > queue.add(aLong > 0?aLong + "个元素入栈":"操作失败"));
5        System.out.println(queue.take());
6        ops.leftPushAll("list - key1","value2","value3","value4")
7            .subscribe( aLong - > queue.add(
8              aLong > 0?aLong + "个元素入栈":"操作失败"));
9        System.out.println(queue.take());
10       ops.leftPop("list - key1").subscribe(
11           aString - > queue.add(aString!= null?aString + "出栈":"操作失败"));
12       System.out.println(queue.take());
13   }
```

如文件 6-9 所示,所有的插入和删除操作均在表头执行,因此可认为 list-key1 是一个栈。第 3 行调用 leftPush()方法将 value1 入栈。该方法的返回值为 reactor. core. publisher. Mono < Long >。Long 类型参数表示执行入栈操作后当前栈内元素的个数。随后调用 Mono 类的 subscribe()方法执行订阅。第 4 行采用 Lambda 表达式对 leftPush()方法的返回值进行处理,生成操作结果的返回消息。其中 queue 是 java. util. concurrent. LinkedBlockingDeque 类型的对象,此处用于存储返回消息。该对象可在@Before 注解标注的方法(@Before 注解是 JUnit 4.0 测试框架提供的注解,用于在每个标注为@Test 的方法执行之前执行)中被实例化,代码如下:

```
LinkedBlockingDeque < String > queue = new LinkedBlockingDeque <>();
```

第 6 行调用 leftPushAll()方法实现元素批量入栈,该方法的返回值类型、含义与 leftPush()方法完全一致。第 10 行调用 leftPop()方法将列表中的第一个元素移除并返回。该方法的返回值为 reactor. core. publisher. Mono < V >,其中的泛型 V 与 ReactiveListOperations < K,V >接口中定义的泛型 V 一致(泛型 V 在 6.3.2 节开头已定义为 String)。在调用了 leftPop()方法后,第 10 行调用 Mono 类的 subscribe()方法执行订阅。该测试用例的运行结果如图 6-4 所示。

```
1个元素入栈
4个元素入栈
value4出栈
```

图 6-4 例 6-5 的运行结果

【例 6-6】 在存放有值 value2、value2、value3、value4 的列表 list-key2 中移除两个 value2。测试代码如文件 6-10 所示。

【文件 6-10】 例 6-6 测试代码

```
1    @Test
2    public void testRemoveOperatioin() {
3      ops.rightPushAll("list-key2","value2","value2","value3",
4          "value4").subscribe();
5      ops.remove("list-key2",2,"value2").subscribe();
6      ops.range("list-key2",0,-1).subscribe(System.out::println);
7      System.out.println("移除 2 个 value2 后，剩余：");
8    }
```

如文件 6-10 所示，第 5 行调用 ReactiveListOperations< K,V >接口的 remove()方法，移除两个值为 value2 的元素。该方法的返回值类型为 reactor. core. publisher. Mono < Long >。其中的 Long 类型参数表示成功删除的元素的个数。第 6 行调用 range()方法输出列表中剩余的全部元素的值。

2. 指定范围取值

用于在 Redis 列表中获取指定范围内的元素的响应式方法有：

reactor. core. publisher. Flux < V > range(K key,long start,long end)：从列表 key 中获取[start,end]区间的元素。列表中的元素索引从 0 开始，start=0,end=-1 表示获取列表中的所有元素。

【例 6-7】 将值 value1～value4 添加到列表 list-key2 的尾部（右侧），并取出列表 list-key2 中的所有元素。测试代码如文件 6-11 所示。

【文件 6-11】 例 6-7 测试代码

```
1    @Test
2    public void testRangeOperation() throws InterruptedException {
3      ops.rightPushAll("list-key2","value1","value2","value3","value4")
4      .subscribe(aLong -> queue.add(aLong > 0?"入栈成功":"入栈失败"));
5      //获取指定范围内的值
6      ops.range("list-key2", 0, -1).subscribe(System.out::println);
7      System.out.println(queue.take() + " 列表元素为：");
8    }
```

如文件 6-11 所示，第 3 行调用 ReactiveListOperations < K,V >接口的 rightPushAll()方法向列表 list-key2 添加元素。该方法的返回值类型为 reactor. core. publisher. Mono < Long >，其中的泛型 Long 表示执行添加操作后当前列表中的元素个数。第 4 行调用 Mono 类的 subscribe()方法订阅操作完成的消息并处理操作结果。第 6 行调用 ReactiveListOperations < K,V >接口的 range()方法，获取当前列表中的所有元素。该方法的返回值类型为 reactor. core. publisher. Flux < V >，其中的泛型 V 的类型与 ReactiveListOperations < K,V > 接口中泛型 V 的类型一致，本节已将泛型 V 定义为 String。随后，调用 Flux 类的 subscribe()方法，用方法引用将列表中的 String 类型元素全部输出。

3. 设置、获取指定下标值

用于在 Redis 列表中设置、获取指定下标值的响应式方法有：

（1）reactor. core. publisher. Mono < Boolean > set(K key,long index,V value)：将列表 key 中索引 index 位置处的值设置为 value。

（2）reactor. core. publisher. Mono < V > index(K key,long index)：获取列表 key 中索引 index 处的值。

【例 6-8】 将值 value1～value4 添加到列表 list-key2 的尾部(右侧)，设置索引为 0 的元素的值为 value2，并取出索引为 0 的元素的值。测试代码如文件 6-12 所示。

【文件 6-12】 例 6-8 测试代码

```
1   @Test
2   public void testIndexOperations() {
3     ops.rightPushAll("list-key2","value1","value2","value3","value4")
4      .subscribe();
5     ops.set("list-key2", 0, "value2")
6      .subscribe(aBool -> queue.add(aBool?"设置成功,新值为: ":"设置失败"));
7     ops.index("list-key2",0).subscribe(System.out::println);
8     queue.forEach(System.out::print);
9   }
```

如文件 6-12 所示，第 5 行调用 ReactiveListOperations < K，V >接口的 set()方法设置索引为 0 的元素的值。该方法的返回值类型为 reactor. core. publisher. Mono < Boolean >。随后，在第 6 行调用 Mono 类的 subscribe()方法完成订阅，并在 Mono 发布操作完成的信号(布尔型)后，由 Lambda 表达式将处理结果转换为输出消息存放到 LinkedBlockingDeque 中。接下来，第 7 行调用 index()方法取出索引为 0 的元素的新值。该方法的返回值类型为 reactor. core. publisher. Mono < V >，其中泛型 V 与 ReactiveListOperations < K，V >接口中的泛型 V 类型一致，这里已将泛型 V 统一定义为 String。随后调用 Mono 类的 subscribe()方法完成订阅。在接收到 Mono 对象发出的操作完成信号后，将操作结果输出到控制台。第 8 行将 LinkedBlockingDeque 对象中存放的消息逐一输出到控制台。

4. 裁剪列表

用于在 Redis 列表中执行裁剪的响应式方法为：

reactor. core. publisher. Mono < Boolean > trim(K key,long start,long end)：保留列表 key 中索引位于[start,end]区间内的元素，删除其他元素。列表中元素的索引从 0 开始。

【例 6-9】 在存放有值 value1、value2、value3、value4 的列表 list-key2 中，保留索引为 1、2 的元素，其余移除。测试代码如文件 6-13 所示。

【文件 6-13】 例 6-9 测试代码

```
1   @Test
2   public void testTrimOperation() {
3     ops.rightPushAll("list-key2","value1","value2","value3","value4")
4        .subscribe();
5     ops.trim("list-key2",1,2).subscribe();
6     ops.range("list-key2",0, -1).subscribe(System.out::println);
7     System.out.println("下标 1,2 保留,其余删除,剩余: ");
8   }
```

如文件 6-13 所示，与例 6-6 中的 remove()方法类似，为实现列表裁剪，第 5 行调用了 ReactiveListOperations < K，V >接口的 trim()方法。该方法的返回值类型为 reactor. core. publisher. Mono < Boolean >。随后，第 6 行调用 range()方法将列表中剩余元素输出到控制台。

6.3.3 操作哈希类型的数据

在操作哈希类型的数据时，搞清楚 Redis 中的哈希类型的结构是至关重要的。图 6-5 是一个存储在 Redis 中的哈希类型数据的例子。

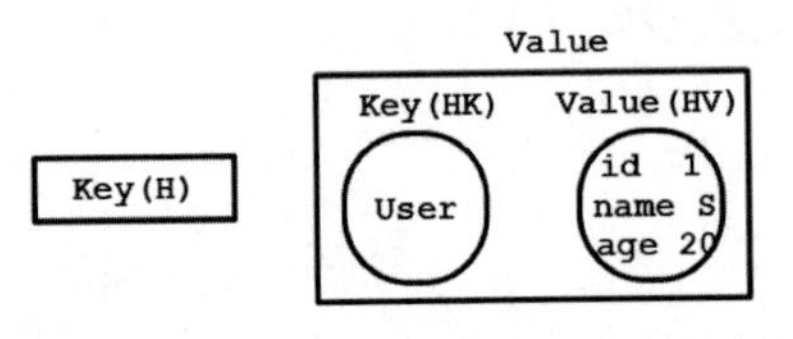

图 6-5 一个 Redis 中哈希类型数据的例子

如图 6-5 所示，Redis 本身是一个键值型数据库。图中的 Key(H)就是数据存放在 Redis 中的键，而 Value 部分就是值。在存放哈希类型数据时，值(Value)部分也是一个哈希类型，分为字段和值两部分，分别以 Key（HK）和 Value（HV）表示。

Spring Data Redis 中，操作哈希类型的响应式接口为 ReactiveHashOperations < H，HK，HV >，该接口定义的方法和 HashOperations、BoundHashOperations 接口定义的方法类似。

可以通过 ReactiveRedisTemplate（或 ReactiveStringRedisTemplate）类的 opsForHash() 方法获取 ReactiveHashOperations 对象，代码如下：

```
ReactiveHashOperations<String,String,String> ops =
    redisTemplate.opsForHash();
```

其中，第一个 String 约定了 Redis 的键(Key(H))的类型，第二个 String 则约定了 Redis 中存放的哈希类型数据的字段(HK)的类型，第三个则约定了哈希类型数据的值(HV)的类型。ReactiveHashOperations 接口的主要方法如下。

1. 添加哈希数据

向 Redis 中添加哈希数据的响应式方法有：

（1）reactor.core.publisher.Mono < Boolean > put（H key，HK hashKey，HV value）：将字段为 hashKey、值为 value 的哈希数据以 key 为键存入 Redis。

（2）reactor.core.publisher.Mono < Boolean > putAll（H key，Map <? extends HK，? extends HV > map）：使用 map 中提供的数据设置多个哈希的字段和值。

（3）reactor.core.publisher.Mono < Boolean > putIfAbsent（H key，HK hashKey，HV value）：仅当字段 hashKey 不存在时设置字段 hashKey 的值 value。

2. 获取哈希值

用于获取哈希值的响应式方法有：

（1）reactor.core.publisher.Mono < HV > get（H key，Object hashKey）：获取键 key 对应哈希中字段 hashKey 的值。

（2）reactor.core.publisher.Flux < Map.Entry < HK，HV >> scan（H key）：使用 Flux 在键 key 处遍历哈希中的条目。

（3）reactor.core.publisher.Mono < List < HV >> multiGet（H key，Collection < HK > hashKeys）：从哈希 key 中获取给定字段 hashKeys 的值。

【例 6-10】 以 hash-key 为键在 Redis 中存储一个哈希，该哈希的字段为 k1，值为 v1，记为< k1，v1 >。随后，再以 hash-key 为键，存入两个哈希：< k2，v2 >和< k3，v3 >。在 hash-key 中不存在< k4，v4 >的情况下将该哈希存入 Redis。随后，分别取出 hash-key 中字段 k1 对应的值和 hash-key 中的全体哈希并输出到控制台。测试代码如文件 6-14 所示。

【文件 6-14】 例 6-10 测试代码

```
1   @Test
2   public void testPutAndGetOperations(){
3     ops.put("hash-key","k1","v1").subscribe();
4     Map<String,String> map = new HashMap<>();
5     map.put("k2","v2");
6     map.put("k3","v3");
7     ops.putAll("hash-key",map).subscribe();
8     ops.putIfAbsent("hash-key","k4","v4").subscribe();
9     ops.get("hash-key","k1").subscribe(str -> System.out.println(
10      "k1 = " + str));
11    Mono<List<Map.Entry<String, String>>> monoEntry = ops.scan(
12      "hash-key").collectList();
13    monoEntry.subscribe(list -> list.forEach(e -> System.out.println(
14      e.getKey() + " = " + e.getValue())));
15  }
```

如文件 6-14 所示，第 3 行调用 ReactiveHashOperations 接口的 put()方法，指定 Redis 的键为 hash-key 和哈希数据< k1，v1 >作为 put()方法的参数。该方法的返回值类型为 reactor.core.publisher.Mono < Boolean >，因此调用 Mono 类的 subscribe()方法完成订阅。如果需要针对 put()方法执行结果做出判定，可参照例 6-4 和例 6-8 的做法，在 subscribe()方法中处理 Mono 对象发布的操作结果。第 7 行调用 putAll()方法，将 HashMap 中的两组哈希< k2，v2 >和< k3，v3 >存入 Redis。随后，第 8 行调用 putIfAbsent()方法将哈希数据< k4，v4 >存入 Redis。第 9 行调用 get()方法获取键 hash-key 对应的哈希数据 k1 的值。该方法的返回值类型为 reactor.core.publisher.Mono < HV >，其中的泛型 HV 与 ReactiveHashOperations < H，HK，HV >接口中的泛型 HV 一致。本节中已将 HV 具体化为 String 类型。因此，在第 9 行调用 subscribe()方法时，会接收 Mono < String >发布的一个 String 类型的返回值 str，subscribe()方法用 Lambda 表达式处理该返回值 str。第 11 行调用 scan()方法，对 hash-key 中的哈希数据进行遍历。该方法的返回值类型为 reactor.core.publisher.Flux < Map.Entry < HK，HV >>。由于 Flux 对象代表可以包含 N 个元素的响应式流，因此调用 Flux 对象的 collectList()方法将此 Flux 对象发布的所有元素收集到一个 java.util.List 中，该 List 在序列完成时由生成的 Mono 对象发布。如果序列为空，则发射空 List。图 6-6 是 collectList()方法的工作原理示意图。第 13 行通过调用 subscribe()方法获取 Mono 对象发布的 List 响应流。随后分别调用 Map.Entry < K，V >接口的 getKey()和 getValue()方法分别获取哈希数据的键和值。

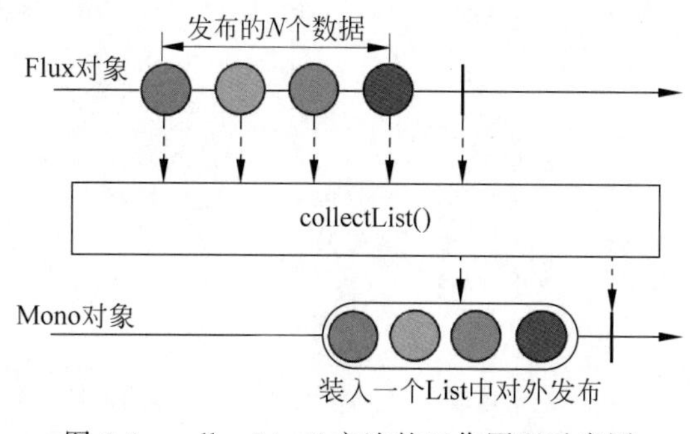

图 6-6 collectList()方法的工作原理示意图

【例 6-11】 获取例 6-10 中添加的所有哈希的值。测试代码如文件 6-15 所示。

【文件 6-15】 例 6-11 测试代码

```
1    @Test
2    public void testGetMultiKeyOperation() {
3        ops.multiGet("hash-key",List.of("k1","k2","k3","k4"))
4                .subscribe(list -> list.forEach(System.out::println));
5    }
```

如文件 6-15 所示，为获取键 hash-key 中所有的哈希的值，需要调用 ReactiveHashOperations < H，HK，HV >接口的 multiGet()方法（第 3 行）。该方法的返回值类型为 reactor. core. publisher. Mono < java. util. List < HV >>，其中的泛型 HV 指定的是 Redis 中存储的哈希的值的类型（见图 6-5），这个泛型与 ReactiveHashOperations < H，HK，HV >接口中定义的泛型一致，本节已将 HV 定义为 String。因此，multiGet()方法的返回值类型为 List < String >。第 4 行调用 Mono 类的 subscribe()方法完成订阅，并采用 Lambda 表达式对 Mono 对象发布的 List < String >执行遍历。

3. 对哈希值的其他操作

对哈希值进行的其他响应式操作还有：

（1）reactor. core. publisher. Mono < Boolean > hasKey(H key，Object hashKey)：确定字段 hashKey 是否存在。

（2）reactor. core. publisher. Mono < Double > increment(H key，HK hashKey，double delta)：按给定增量 delta 增加哈希 key 中的字段 hashKey 对应的值。

（3）reactor. core. publisher. Mono < Long > increment(H key，HK hashKey，long delta)：按给定增量 delta 增加哈希 key 中的字段 hashKey 对应的值。

【例 6-12】 判断 hash-key 中是否存在一个哈希 k1，在控制台输出判断结果。随后，再以 hash-key 为键，存入两个哈希：< nk1，100 >和< nk2，300 >。将 nk1 和 nk2 的值分别增加 50.0（Double 类型）和 100L（Long 类型），输出增加后的哈希值。测试代码如文件 6-16 所示。

【文件 6-16】 例 6-12 测试代码

```
1    @Test
2    public void testHasKeyOperations() {
3      ops.hasKey("hash-key","k1")
4        .subscribe(aBool -> System.out.println(aBool?"k1 存在":"k1 不存在"));
5    }
6
7    @Test
8    public void testIncrementOperations(){
9      ops.put("hash-key","nk1","100").subscribe();
10     ops.increment("hash-key","nk1",50.0)
11      .subscribe(aDouble -> System.out.println("the result is " + aDouble));
12     ops.put("hash-key","nk2","300").subscribe();
13     ops.increment("hash-key","nk2",100L)
14        .subscribe(aLong -> System.out.println("the result is " + aLong));
15   }
```

4. 根据键操作哈希

根据键操作哈希的响应式方法有：

(1) reactor. core. publisher. Flux < Map. Entry < HK,HV >> entries(H key)：获取指定键 key 存储的哈希。

(2) reactor. core. publisher. Flux < HK > keys(H key)：获取键 key 对应的哈希中所有的字段。

(3) reactor. core. publisher. Flux < HV > values(H key)：获取键 key 对应的哈希中所有的值。

(4) reactor. core. publisher. Mono < Long > size(H key)：获取键 key 对应的哈希的大小。

【例 6-13】 对例 6-10 中的哈希，分别获取键 hash-key 对应的所有哈希，在控制台输出所有的哈希键及所有的哈希值。测试代码如文件 6-17 所示。

【文件 6-17】 例 6-13 测试代码

要求 1：获取键 hash-key 对应的所有哈希。

```
1   @Test
2   public void testMapEntryOperations() {
3       //准备数据部分见例 6-10
4       Mono<List<Map.Entry<String,String>>> monoEnts =
5           ops.entries("hash-key").collectList();
6       monoEnts.subscribe(list -> list.forEach(entry ->
7           System.out.println(entry.getKey() + " = " + entry.getValue())));
8   }
```

要获取 Redis 键对应的所有哈希，可调用 ReactiveHashOperations < H,HK,HV >接口的 entries()方法。该方法的返回值类型为 reactor. core. publisher. Flux < java. util. Map. Entry < HK,HV >>(如第 4、5 行所示)。与例 6-10(文件 6-14)的处理方式类似，第 5 行调用了 Flux 对象的 collectList()方法，将 Flux 对象转换为 Mono 对象，原 Flux 对象发布的多个元素存放在一个 List 中由新的 Mono 对象集中发布。而 collectList()方法的返回值类型为 reactor. core. publisher. Mono < java. util. List < T >>，此处 List 中的泛型 T 与第 4、5 行调用 entries()方法时生成的 Flux 对象的泛型一致，为 java. util. Map. Entry < HK,HV >。因此，第 6 行的元素 list 的类型为 java. util. List < java. util. Map. Entry < String,String >>。在对该 list 进行遍历时，得到的每个元素 entry 的类型即为 java. util. Map. Entry < String,String >。随后，在第 7 行调用 java. util. Map. Entry < String,String >接口的 getKey()和 getValue()方法分别获取哈希字段和哈希值。

要求 2：获取键 hash-key 对应的所有哈希字段。

```
1   @Test
2   public void testMapEntryKeyOperations() {
3       Mono<List<String>> monoList = ops.keys("hash-key").collectList();
4       monoList.subscribe(list -> list.forEach(System.out::println));
5   }
```

要求3：获取键hash-key对应的所有哈希值。

```
1    @Test
2    public void testMapEntryValueOperations() {
3        ops.values("hash - key").collectList()
4                .subscribe(list -> list.forEach(System.out::println));
5    }
```

5. 删除操作

对哈希数据执行删除的响应式方法有：

（1）reactor.core.publisher.Mono<Boolean> delete(H key)：移除给定的键key。

（2）reactor.core.publisher.Mono<Long> remove(H key,Object...hashKeys)：从键key对应的哈希中删除给定的哈希字段hashKeys。

【例6-14】 对于例6-10中的哈希删除键hash-key对应的哈希<k1,v1>，计算删除该哈希后键hash-key对应的哈希大小，最后删除键hash-key。测试代码如文件6-18所示。

【文件6-18】 例6-14测试代码

```
1    @Test
2    public void testDeleteOperations(){
3      ops.remove("hash - key","k1").subscribe(aLong ->
4        System.out.println(aLong > 0?"k1删除成功":"k1删除失败"));
5      ops.size("hash - key").subscribe(aLong -> System.out.println(
6        "size is " + aLong));
7      ops.delete(("hash - key")).subscribe(aBoolean ->
8        System.out.println(aBoolean?"hash - key已删除":"hash - key删除失败"));
9    }
```

运行此测试代码，控制台输出如图6-7所示（数据准备可见例6-10）。

```
现有数据：
k1=v1
k3=v3
k2=v2
k4=v4
k1删除成功
size is 3
hash-key已删除
```

图6-7 例6-14测试代码运行结果

6.3.4 操作集合类型的数据

Spring Data Redis中，操作集合类型的响应式接口为ReactiveSetOperations<K,V>，该接口定义的方法和SetOperations、BoundSetOperations接口定义的方法类似。

可以通过ReactiveRedisTemplate（或ReactiveStringRedisTemplate）类的opsForSet()方法获取ReactiveSetOperations对象，代码如下：

```
ReactiveSetOperations < String, String > ops =  reactiveRedisTemplate.opsForSet();
```

其中，ReactiveSetOperations接口的主要方法如下。

1. 向集合中添加元素

reactor. core. publisher. Mono < Long > add(K key,V...values)：将给定的元素 values 添加到键 key 对应的集合。

2. 返回所有元素

reactor. core. publisher. Flux < V > members(K key)：返回指定键 key 对应集合的所有元素。

3. 计算集合大小

reactor. core. publisher. Mono < Long > size(K key)：返回键 key 对应集合的大小。

4. 判断元素是否为集合的成员

reactor. core. publisher. Mono < Boolean > isMember(K key,Object o)：检查指定的值 o 是否是键 key 对应集合中的成员。如果是，则返回 true；如果不是，则返回 false。

【例 6-15】 向键 set-key 对应的集合中添加元素 value1、value2 和 value3，计算添加元素后的集合的大小，在控制台输出集合的所有元素，并判断 value2 是否为本集合的元素。测试代码如文件 6-19 所示。

【文件 6-19】 例 6-15 测试代码

```
1   @Test
2   public void testSetOperations1(){
3     ops.add("set - key","value1","value2","value3").subscribe();
4     ops.size("set - key").subscribe(
5       aLong  - > System.out.println("The size of set - key is : " + aLong));
6     ops.members("set - key").collectList().subscribe(
7       list - > list.forEach(e - > System.out.println("The members : " + e)));
8     ops.isMember("set - key","value2").subscribe(
9       aBoolean  - > System.out.println("value2 :" + aBoolean));
10  }
```

5. 从集合中随机抽取多个元素

从集合中随机抽取多个元素的响应式方法有：

(1) reactor. core. publisher. Mono < V > randomMember(K key)：从键 key 对应的集合中随机返回一个元素。

(2) reactor. core. publisher. Flux < V > randomMembers(K key,long count)：从键 key 对应的集合中随机返回 count 个元素。

(3) reactor. core. publisher. Flux < V > distinctRandomMembers(K key,long count)：随机从键 key 对应的集合中获取多个元素，这些元素不能相同。

6. 移除元素

从集合中移除(弹出)元素的响应式方法有：

(1) reactor. core. publisher. Mono < V > pop(K key)：从键 key 对应的集合中随机移除且返回一个元素。

(2) reactor. core. publisher. Flux < V > pop(K key,long count)：从键 key 对应集合中随机移除且返回 count 个元素。

(3) reactor. core. publisher. Mono < Long > remove(K key,Object...values)：从键 key

对应的集合中移除值为 values 的元素。

7. 删除键

reactor. core. publisher. Mono<Boolean>delete(K key)：删除给定的键 key。

【例 6-16】 向键 set-key 对应的集合中添加元素 value1、value2、value3 和 value4，移除值为 value4 的元素，随后从剩余元素中随机移除两个，再随机抽取两个元素，最后删除集合 set-key。测试代码如文件 6-20 所示。

【文件 6-20】 例 6-16 测试代码

```
1   @Test
2   public void testSetOperations2(){
3   ops.add("set-key","value1","value2","value3","value4").subscribe();
4     ops.remove("set-key","value4").subscribe();
5     ops.members("set-key").collectList().subscribe(
6       list -> list.forEach(e-> System.out.println("移除 4,现有元素: "+e)));
7     ops.randomMembers("set-key",2).collectList().subscribe(
8       list -> list.forEach(e-> System.out.println("随机取出两个元素: "+e)));
9     ops.pop("set-key",2).collectList().subscribe(
10      list-> list.forEach(e-> System.out.println("随机弹出两个: "+e)));
11    ops.members("set-key").collectList().subscribe(
12      list -> list.forEach(e-> System.out.println("剩余元素: "+e)));
13    ops.delete("set-key").subscribe(aBoolean -> System.out.println(
14      aBoolean?"set-key 删除成功":"set-key 删除失败"));
15  }
```

```
移除4，现有元素： value3
移除4，现有元素： value2
移除4，现有元素： value1
随机取出两个元素：value1
随机取出两个元素：value3
随机弹出两个：value1
随机弹出两个：value3
剩余元素： value2
set-key删除成功
```

图 6-8 例 6-16 测试代码运行结果

该测试代码的运行结果如图 6-8 所示。

8. 求并集

求集合并集的响应式方法有：

(1) reactor. core. publisher. Flux<V>union(K key,Collection<K>otherKeys)：合并 key 集合和 otherKeys 集合中的元素。

(2) reactor. core. publisher. Flux<V>union(K key,K otherKey)：合并 key 集合和 otherKey 集合中的元素。

(3) reactor. core. publisher. Mono<Long>unionAndStore(K key,Collection<K>otherKeys,K destKey)：合并 key 集合和 otherKeys 集合中的元素，并且将合并结果存储到 destKey 集合。

(4) reactor. core. publisher. Mono<Long>unionAndStore(K key,K otherKey,K destKey)：合并 key 集合和 otherKeys 集合中的元素，并且将合并结果存储到 destKey 集合。

【例 6-17】 集合 A={v1,v2,v3}，集合 B={v1,v4,v5}，集合 C={v4,v6,v7}。求集合 A、B、C 的并集。测试代码如文件 6-21 所示。

【文件 6-21】 例 6-17 测试代码

```
1   @Test
2   public void testUnionOperation() {
3       ops.add("A","v1","v2","v3").subscribe();
4       ops.add("B","v1","v4","v5").subscribe();
5       ops.add("C","v4","v6","v7").subscribe();
```

```
6        System.out.println("并集中的元素为：");
7        ops.union("A", List.of("B","C")).collectList().subscribe(
8          list -> list.forEach(e -> System.out.print(e + " ")));
9    }
```

测试代码运行结果如图 6-9 所示。

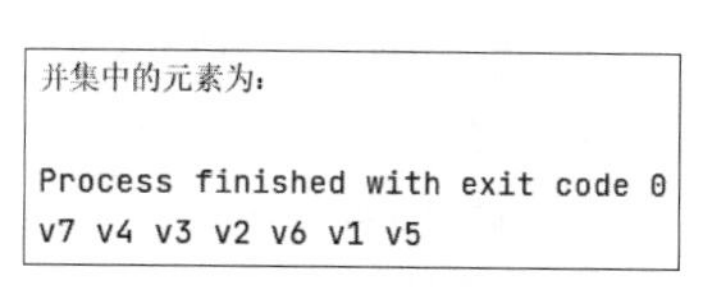

图 6-9 例 6-17 测试代码运行结果

9. 求交集

求集合交集的响应式方法有：

（1）reactor. core. publisher. Flux < V > intersect(K key,Collection < K > otherKeys)：计算 key 集合和 otherKeys 集合所有元素的交集。

（2）reactor. core. publisher. Flux < V > intersect(K key,K otherKey)：计算 key 集合和 otherKey 集合所有元素的交集。

（3）reactor. core. publisher. Mono < Long > intersectAndStore(K key,Collection < K > otherKeys,K destKey)：计算 key 集合和 otherKeys 集合所有元素的交集，并且将结果存储到 destKey 集合。

（4）reactor. core. publisher. Mono < Long > intersectAndStore(K key,K otherKey,K destKey)：计算 key 集合和 otherKey 集合所有元素的交集，并且将结果存储到 destKey 集合。

【例 6-18】 集合 A={v1,v2,v3}，集合 B={v1,v2,v4}，集合 C={v2,v3,v4}。求集合 A、B、C 的交集 D。测试代码如文件 6-22 所示。

【文件 6-22】 例 6-18 测试代码

```
1    @Test
2    public void testIntersectOperation() {
3      ops.add("A","v1","v2","v3").subscribe();
4      ops.add("B","v1","v2","v4").subscribe();
5      ops.add("C","v2","v3","v4").subscribe();
6      ops.intersectAndStore("A",List.of("B","C"),"D")
7         .subscribe(aLong -> System.out.println("集合 D 有" + aLong + "个元素"));
8      ops.members("D").collectList().subscribe(
9         list -> list.forEach(System.out::print));
10   }
```

```
集合D有 1 个元素

Process finished with exit code 0
v2
```

图 6-10 例 6-18 测试代码运行结果

测试代码运行结果如图 6-10 所示。

10. 求差集

求集合差集的响应式方法有：

（1）reactor. core. publisher. Flux < V > difference(K key,Collection < K > otherKeys)：计算 key 集合和 otherKeys 对应集合的差集，返回 key 集合中没有在 otherKeys 集合中存在的元素列表。

（2）reactor. core. publisher. Flux < V > difference(K key,K otherKey)：计算 key 集合和 otherKey 集合的差集，返回 key 集合中没有在 otherKey 集合中存在的元素列表。

（3）reactor. core. publisher. Mono < Long > differenceAndStore(K key,Collection < K >

otherKeys,K destKey)：计算 key 集合和 otherKeys 对应集合的差集，并且将 key 集合中没有在 otherKeys 集合中存在的元素列表存储到 destKey 集合。

（4）reactor.core.publisher.Mono <Long> differenceAndStore(K key,K otherKey,K destKey)：计算 key 和 otherKey 对应集合的差集，并且将 key 集合中没有在 otherKey 集合中存在的元素列表存储到 destKey 集合。

【例 6-19】 集合 A={v1,v2,v3}，集合 B={v1,v2,v4}。求集合 A 与集合 B 的差集。测试代码如文件 6-23 所示。

【文件 6-23】 例 6-19 测试代码

```
1   @Test
2   public void testDifferenceOperation() {
3     ops.add("A","v1","v2","v3").subscribe();
4     ops.add("B","v1","v2","v4").subscribe();
5     ops.difference("A","B").collectList().subscribe(
6       list -> list.forEach(System.out::println));
7   }
```

11. 移动元素

reactor.core.publisher.Mono <Boolean> move(K sourceKey,V value,K destKey)：将指定 value 元素从 sourceKey 集合移动到 destKey 集合。

【例 6-20】 集合 A={v1,v2,v3,v4}，将集合 A 中的元素 v1 存放到集合 B 中。测试代码如文件 6-24 所示。

【文件 6-24】 例 6-20 测试代码

```
1   @Test
2   public void testMovingOperation() {
3     ops.add("A","v1","v2","v3","v4").subscribe();
4     ops.move("A","v1","B").subscribe();
5     ops.members("B").collectList().subscribe(
6       list -> list.forEach(System.out::println));
7   }
```

6.3.5 操作有序集合类型的数据

Spring Data Redis 中，操作有序集合类型的响应式接口为 ReactiveZSetOperations <K,V>，该接口定义的方法和 ZSetOperations、BoundZSetOperations 接口定义的方法类似。

可以通过 ReactiveRedisTemplate（或 ReactiveStringRedisTemplate）类的 opsForZSet() 方法获取 ReactiveZSetOperations 对象，代码如下：

```
ReactiveZSetOperations<String,String> ops = reactiveRedisTemplate.opsForZSet();
```

其中，ReactiveZSetOperations 接口的主要方法如下。

1. 添加元素

向有序集合中添加元素的响应式方法有：

（1）reactor.core.publisher.Mono <Boolean> add(K key,V value,double score)：将

元素 value 添加到键 key 对应的有序集合，如果它已经存在则更新其分数 score。

（2） reactor. core. publisher. Mono < Long > addAll（K key，Collection <? extends ZSetOperations. TypedTuple < V >> tuples）：将 tuples 添加到键 key 对应的有序集合中，如果它们已经存在则更新其分数。

2. 获取集合大小

获取集合大小的响应式方法有：

（1） reactor. core. publisher. Mono < Long > size（K key）：返回键 key 对应的有序集合元素的个数。

（2） reactor. corc. publisher. Mono < Long > count（K key，org. springframework. data. domain. Range < Double > range）：计算 key 对应的有序集合中分数在最小值（min）和最大值（max）之间的元素个数。注：本节中用到的 Range 类均指 org. springframework. data. domain. Range < T >，此后不再赘述。

3. 获取分数

reactor. core. publisher. Mono < Double > score（K key，Object o）：使用键 key 从有序集合获取值为 o 的元素的分数（score）。

4. 返回值的索引

（1） reactor. core. publisher. Mono < Long > rank（K key，Object o）：确定值 o 在有序集中的索引。

（2） reactor. core. publisher. Mono < Long > reverseRank（K key，Object o）：当得分从高到低排序时，确定排序集中具有值 o 的元素的索引。

5. 遍历集合

reactor. core. publisher. Flux < ZSetOperations. TypedTuple < V >> scan（K key）：使用 Flux 对象遍历 key 集合。

【例 6-21】 元素 v1、v2 和 v3 的分数分别为 24.0、21.0 和 22.0。完成以下操作：将 3 个元素放入有序集合 Z 中，查看放入元素后 Z 集合的元素个数，查看元素 v2 的分数及排名（分数由低到高排名），统计分数范围在[21.0，24.0）内的元素个数。测试代码如文件 6-25 所示。

【文件 6-25】 例 6-21 测试代码

```
1    @Test
2    public void testZSetOperations1() {
3      ZSetOperations.TypedTuple<String> dtt1 = new
4         DefaultTypedTuple<>("v1",24.0);
5      ZSetOperations.TypedTuple<String> dtt2 = new
6         DefaultTypedTuple<>("v2",21.0);
7      ZSetOperations.TypedTuple<String> dtt3 = new
8         DefaultTypedTuple<>("v3",22.0);
9      ops.addAll("Z", List.of(dtt1,dtt2,dtt3)).subscribe();
10     ops.size("Z").subscribe(System.out::println);
11     ops.score("Z","v2").subscribe(System.out::println);
12     ops.rank("Z","v2").subscribe(System.out::println);
13     ops.count("Z", Range.of(Range.Bound.inclusive(21.0),
14        Range.Bound.exclusive(24.0))).subscribe(System.out::println);
15   }
```

6. 分数递增

reactor. core. publisher. Mono < Double > incrementScore(K key, V value, double delta)：按增量 delta 递增有序集合中值为 value 的元素的分数。

7. 删除指定键

reactor. core. publisher. Mono < Boolean > delete(K key)：移除给定的键 key。

8. 移除元素

（1）reactor. core. publisher. Mono < Long > remove(K key, Object... values)：从有序集合中删除值。

（2）reactor. core. publisher. Mono < Long > removeRange(K key, Range < Long > range)：从键 key 对应的有序集合中删除指定范围内的元素。

（3）reactor. core. publisher. Mono < Long > removeRangeByScore(K key, Range < Double > range)：从键 key 对应的有序集合中删除分数在 min 和 max 之间的元素。

【例 6-22】 元素 v1、v2 和 v3 的分数分别为 24.0、21.0 和 22.0。完成以下操作：将 3 个元素放入有序集合 Z 中，将元素 v1 的分数增加 12，删除下标 0、1 的元素，最后删除集合 Z。测试代码如文件 6-26 所示。

【文件 6-26】 例 6-22 测试代码

```
1    @Test
2    public void testZSetOperations2() {
3        ZSetOperations.TypedTuple < String > dtt1 = new
4          DefaultTypedTuple <>("v1",24.0);
5        ZSetOperations.TypedTuple < String > dtt2 = new
6          DefaultTypedTuple <>("v2",21.0);
7        ZSetOperations.TypedTuple < String > dtt3 = new
8          DefaultTypedTuple <>("v3",22.0);
9        ops.addAll("Z",List.of(dtt1,dtt2,dtt3)).subscribe();
10       ops.incrementScore("Z","v1",12.0).subscribe(
11         aDouble -> System.out.println("new value of v1 is " + aDouble));
12       ops.removeRange("Z",Range.of(Range.Bound.inclusive(0L),
13         Range.Bound.exclusive(1L))).subscribe();
14       ops.delete("Z").subscribe();
15   }
```

9. 范围操作

对有序集合在指定范围内执行操作的响应式方法有：

（1）reactor. core. publisher. Flux < V > range(K key, Range < Long > range)：从有序集合中获取由 range 指定的开始(start)和结束(end)之间的元素。

（2）reactor. core. publisher. Flux < V > rangeByScore(K key, Range < Double > range)：从有序集合中获取分数在 range 指定的最小值(min)和最大值(max)之间的元素。

（3）reactor. core. publisher. Flux < V > rangeByScore(K key, Range < Double > range, RedisZSetCommands. Limit limit)：从有序集合中获取从开始到结束的元素，其中分数介于最小值和最大值之间。

（4）reactor. core. publisher. Flux < ZSetOperations. TypedTuple < V >> rangeByScoreWithScores (K key, Range < Double > range)：从有序集合中获取分数介于最小值和最大值之间的

RedisZSetCommands. Tuples 集合。

(5) reactor. core. publisher. Flux < ZSetOperations. TypedTuple < V >> rangeByScoreWithScores (K key, Range < Double > range, RedisZSetCommands. Limit limit): 从开始到结束的范围内获取一组 RedisZSetCommands. Tuples,其中分数在最小值和最大值之间。

(6) reactor. core. publisher. Flux < ZSetOperations. TypedTuple < V >> rangeWithScores (K key, Range < Long > range): 从有序集合中获取开始和结束之间的一组元素(RedisZSetCommands. Tuples)。

(7) reactor. core. publisher. Flux < V > reverseRange(K key, Range < Long > range): 按从高到低的顺序排列有序集合中的元素,然后获取开始到结束之间的元素。

(8) reactor. core. publisher. Flux < ZSetOperations. TypedTuple < V >> reverseRangeWithScores (K key, Range < Long > range): 从由高到低排序的有序集合中,从开始到结束的范围内获取一组元素(RedisZSetCommands. Tuples)。

【例 6-23】 元素 v1、v2、v3、v4 和 v5,它们的分数分别为 51.0、31.0、22.0、12.0 和 47.0。将 5 个元素放入有序集合 Z 中。要求完成以下操作: ①输出下标 0、1 的所有元素; ②输出分数在[10,40]范围内的元素; ③输出分数在[30,∞)范围内的元素; ④输出下标除 0、1 外的元素的值和分数; ⑤对集合中的全部元素按分数从高到低排序,并输出排序后的元素的值和分数。测试代码如文件 6-27 所示。

【文件 6-27】 例 6-23 测试代码

```
1   @Test
2   public void testRangeOperations(){
3     //向 Z 集合中存入 5 个元素及分数,代码略
4     //要求①输出下标 0、1 的所有元素
5     ops.range("Z",Range.of(Range.Bound.inclusive(0L),
6       Range.Bound.inclusive(1L))).collectList().subscribe(
7       list -> list.forEach(System.out::println));
8     //要求②输出分数在[10, 40]范围内的元素
9     ops.rangeByScore("Z",Range.of(Range.Bound.inclusive(10D),
10      Range.Bound.inclusive(40D))).collectList().subscribe(
11      list -> list.forEach(System.out::println));
12    //要求③输出分数在[30, ∞)范围内的元素
13    ops.rangeByScore("Z",Range.rightUnbounded(
14      Range.Bound.inclusive(30D))).collectList().subscribe(
15      list -> list.forEach(System.out::println));
16    //要求④输出下标除 0、1 外的元素的值和分数
17    ops.rangeWithScores("Z",Range.rightUnbounded(
18      Range.Bound.inclusive(2L))).collectList().subscribe(
19      list -> list.forEach(System.out::println));
20    //要求⑤对集合中的全部元素按分数从高到低排序,并输出排序后的元素的值和分数
21    ops.reverseRangeWithScores("Z",Range.of(Range.Bound.inclusive(0L),
22     Range.Bound.inclusive(4L))).collectList().subscribe(
23     list -> list.forEach(System.out::println));  }
```

10. 并集和交集

求解有序集合并集和交集的响应式方法有:

(1) reactor. core. publisher. Mono < Long > unionAndStore(K key, Collection < K >

otherKeys,K destKey)：合并 key 集合和 otherKeys 集合，并且将合并结果存储到 destKey 集合中。

（2）reactor. core. publisher. Mono < Long > unionAndStore（K key, K otherKey, K destKey)：合并 key 集合和 otherKey 集合，并且将合并结果存储到 destKey 集合中。

（3）reactor. core. publisher. Mono < Long > intersectAndStore(K key,Collection < K > otherKeys,K destKey)：计算 key 集合和 otherKeys 对应集合的交集，并将结果存储到 destKey 集合中。

（4）reactor. core. publisher. Mono < Long > intersectAndStore（K key,K otherKey,K destKey)：计算 key 集合和 otherKey 对应集合的交集，并将结果存储到 destKey 集合中。

【例 6-24】 元素 v1、v2、v3 和 v4，它们的分数分别为 24.0、21.0、22.0 和 30.0。将这些元素存入集合 A 和集合 B，A={v1,v2,v3}，B={v1,v2,v4}。求解 C=A∪B，D=A∩B。测试代码如文件 6-28 所示。

【文件 6-28】 例 6-24 测试代码

```
1     @Test
2     public void testUnionOperationsBetweenSets() {
3       //分别向 A、B 集合加入元素，此处略
4         ops.unionAndStore("A","B","C").subscribe();
5         ops.scan("C").collectList().subscribe(typedTuples ->
6             typedTuples.forEach(stringTypedTuple ->
7                   System.out.println(stringTypedTuple.getValue()
8                   + ":" + stringTypedTuple.getScore())));
9     }
```

测试代码运行结果：

```
v3 : 22.0  v4 : 30.0  v2 : 42.0  v1 : 48.0
10     @Test
11     public void testUnionOperationsBetweenSets() {
12       //分别向 A、B 集合加入元素，此处略
13       ops.intersectAndStore("A","B","D").subscribe();
14       ops.scan("D").collectList().subscribe(typedTuples ->
15         typedTuples.forEach(stringTypedTuple ->
16                 System.out.println(stringTypedTuple.getValue()
17                 + ":" + stringTypedTuple.getScore())));
18     }
```

测试代码运行结果：

```
v2 : 42.0   v1 : 48.0
```

6.3.6 操作地理空间类型的数据

Spring Data Redis 中，操作地理空间数据类型的响应式接口为 ReactiveGeoOperations < K,M >，该接口定义的方法和 GeoOperations、BoundGeoOperations 接口定义的方法类似。可以通过 ReactiveRedisTemplate 类的 opsForGeo()方法获取 ReactiveGeoOperations 对象，代码如下：

```
ReactiveGeoOperations<String,String> ops = reactiveRedisTemplate.opsForGeo();
```

其中，ReactiveGeoOperations 接口的主要方法如下。

1. 添加坐标

向 Redis 中添加坐标的响应式方法有：

（1）reactor. core. publisher. Mono < Long > add(K key, Iterable < RedisGeoCommands. GeoLocation < M >> locations)：将 RedisGeoCommands. GeoLocations 添加到 key。

（2）reactor. core. publisher. Mono < Long > add(K key, Map < M, org. springframework. data. geo. Point > memberCoordinateMap)：将 memberCoordinateMap 中的坐标批量添加到 key。

（3）reactor. core. publisher. Mono < Long > add(K key, org. springframework. data. geo. Point point, M member)：将具有给定成员名称的 Point 添加到 key。

（4）reactor. core. publisher. Flux < Long > add(K key, org. reactivestreams. Publisher <? extends Collection < RedisGeoCommands. GeoLocation < M >>> locations)：将 RedisGeoCommands. GeoLocations 添加到 key。

（5）reactor. core. publisher. Mono < Long > add(K key, RedisGeoCommands. GeoLocation < M > location)：将 RedisGeoCommands. GeoLocation 添加到 key。

2. 获取位置点表示

获取某位置的点坐标表示的响应式方法有：

（1）reactor. core. publisher. Mono < List < org. springframework. data. geo. Point >> position(K key, M... members)：获取一个或多个成员位置的点表示。

（2）reactor. core. publisher. Mono < org. springframework. data. geo. Point > position (K key, M member)：获取指定成员位置的点表示。

3. 测距

reactor. core. publisher. Mono < org. springframework. data. geo. Distance > distance(K key, M member1, M member2)：获取 member1 和 member2 之间的距离。

【例 6-25】 3 个地点 WestLake、Shanghai 和 Beijing 的坐标分别为(120.09,30.15)、(120.51,30.41)和(116.28,39.55)。将 WestLake 的地理信息以 geo-key 为键存入 Redis。将 Shanghai 和 Beijing 的地理信息以 geo-key 为键批量存入 Redis。将 WestLake 的地理信息以位置点表示，计算 Shanghai 和 Beijing 之间的直线距离，单位：km。测试代码如文件 6-29 所示。

【文件 6-29】 例 6-25 测试代码

```
1   @Test
2   public void testAddGeoOperation(){
3     Point point = new Point(120.09,30.15);
4     ops.add("geo-key",point,"WestLake").subscribe();
5     Map<String, Point> pointMap = new HashMap<>();
6     pointMap.put("Shanghai",new Point(120.51, 30.41));
7     pointMap.put("Beijing",new Point(116.28,39.55));
8     ops.add("geo-key",pointMap).subscribe();
9     ops.position("geo-key","WestLake")
```

```
10            .subscribe(p -> System.out.println("WestLake Location : " +
11            p.getX() + "," + p.getY()));
12        ops.distance("geo-key","Shanghai","Beijing", Metrics.KILOMETERS)
13            .subscribe(distance -> System.out.println(
14            "Distance between Shanghai and Beijing : " +
15            distance.getValue() + distance.getMetric().getAbbreviation()));
16    }
```

4. 获取 Geo 哈希

获取 Geo 哈希的响应式方法有：

（1）reactor.core.publisher.Mono<List<String>> hash(K key,M...members)：获取一个或多个成员的位置的 Geo 哈希表示。

（2）reactor.core.publisher.Mono<String> hash(K key,M member)：获取指定成员的位置 Geo 哈希表示。

5. 计算范围

利用下述方法可以以响应方式获取以一个点为圆心，计算指定半径圆范围内的地理位置信息：

（1）reactor.core.publisher.Flux<org.springframework.data.geo.GeoResult<RedisGeoCommands.GeoLocation<M>>> radius(K key,org.springframework.data.geo.Circle within)：获取给定圆 within 边界内的成员。

（2）reactor.core.publisher.Flux<org.springframework.data.geo.GeoResult<RedisGeoCommands.GeoLocation<M>>> radius(K key,org.springframework.data.geo.Circle within,RedisGeoCommands.GeoRadiusCommandArgs args)：应用参数 args 获取给定圆 within 边界内的成员。

（3）reactor.core.publisher.Flux<org.springframework.data.geo.GeoResult<RedisGeoCommands.GeoLocation<M>>> radius(K key,M member,org.springframework.data.geo.Distance distance)：获取由成员坐标和给定半径定义的圆内的成员。

（4）reactor.core.publisher.Flux<org.springframework.data.geo.GeoResult<RedisGeoCommands.GeoLocation<M>>> radius(K key,M member,org.springframework.data.geo.Distance distance,RedisGeoCommands.GeoRadiusCommandArgs args)：获取由成员坐标定义的圆内的成员，并应用 Metric 和 RedisGeoCommands.GeoRadiusCommandArgs 给定半径。

（5）reactor.core.publisher.Flux<org.springframework.data.geo.GeoResult<RedisGeoCommands.GeoLocation<M>>> radius(K key,M member,double radius)：获取由成员坐标和给定半径定义的圆内的成员。

6. 删除坐标

reactor.core.publisher.Mono<Long> remove(K key,M...members)：删除指定的一个或多个坐标。

7. 删除键

reactor.core.publisher.Mono<Boolean> delete(K key)：删除给定的键。

【例 6-26】 在例 6-25 中 3 个地点的基础上再添加 3 个地点 LouWaiLou、DaTieGuan

和 LongJingCun，其坐标分别为(120.092，30.152)、(120.089，30.151)和(120.10，30.153)将上述 3 个地点的地理信息以 geo-key 为键存入 Redis。计算 Beijing 的地理信息哈希，并以 WestLake 地点为中心，3km 为半径，查找该范围内的地点。随后，删除 LongJingCun 的坐标。最后，删除键 geo-key。测试代码如文件 6-30 所示。

【文件 6-30】 例 6-26 测试代码

```
1    @Test
2    public void testGeoOperations2() {
3      ops.hash("geo - key","Beijing").subscribe(str ->
4        System.out.println("The geo hash of Beijing is " + str));
5      Circle circle = new Circle(new Point(120.09,30.15),
6        new Distance(3,Metrics.KILOMETERS));
7      ops.radius("geo - key",circle).collectList().subscribe(
8        geoResults -> geoResults.forEach(
9          geoLocationGeoResult -> System.out.println(
10           "Geo Points near WestLake within 3Km are: " +
11           geoLocationGeoResult.getContent().getName())));
12     ops.remove("geo - key","LongJingCun").subscribe();
13     ops.delete("geo - key").subscribe(); }
```

如文件 6-30 所示，第 3 行调用 ReactiveGeoOperations 接口的 hash()方法获取 geo-key 中指定成员的地理位置哈希(Geo 哈希)。第 5、6 行以指定坐标(WestLake 的坐标)为圆心，3km 为半径指定一个搜索范围。第 7 行调用 radius()方法在指定范围内进行搜索。对于 radius()方法，也可以采用下述形式的调用：

```
ops.radius("geo - key","WestLake", new Distance(3,Metrics.KILOMETERS))
```

其中，radius()方法的返回值类型为 Flux < GeoResult < RedisGeoCommands. GeoLocation < M >>>，第 7 行首先调用 Flux 对象的 collecList()方法，该方法的返回值类型为 Mono < java. util. List < T >>，这意味着可以将 Flux 准备发布的多个消息封装到一个 List 对象，由生成的 Mono 对象集中发布。其中，泛型 T 与 Flux 中声明的泛型一致。随后调用 subscribe()方法，获取 Mono 对象发布的 GeoResult 列表 List < GeoResult <>>。随后，用 Lambda 表达式遍历 List 列表中的 GeoResult 对象(第 8～11 行)，调用 GeoResult 对象的 getContent()方法获取其中的 RedisGeoCommands. GeoLocation < M >对象，其中泛型 M 与 ReactiveGeoOperations < K，M >中的泛型 M 一致，本节已将泛型 M 定义为 String。调用 GeoLocation 对象的 getName()方法获取地点的名称(第 11 行)。测试代码的运行结果如图 6-11 所示。

```
The geo hash of Beijing is wx48ypbe2q0
Geo Points near WestLake within 3Km are: WestLake
Geo Points near WestLake within 3Km are: DaTieGuan
Geo Points near WestLake within 3Km are: LouWaiLou
Geo Points near WestLake within 3Km are: LongJingCun
```

图 6-11 例 6-26 测试代码运行结果

6.4 响应式发布-订阅

Spring Data 为 Redis 提供了专用的消息服务，这与 Spring 框架中整合的 JMS 在功能和名称上都十分相似。Redis 消息传递可以分为两个功能域，即消息的生产或发布以及消息的消费或订阅。ReactiveRedisTemplate 类用于消息生成。对于异步接收，Spring Data 提供了一个用于消费消息流的消息侦听器容器。如果只实现消息订阅，可使用 ReactiveRedisTemplate 提供的精简的消息侦听器容器。Redis 消息传递的核心功能由包 org. springframework. data. redis. connection 和 org. springfframework. data. redis. listener 提供。

6.4.1 响应式消息发布

要发布消息，可以使用 ReactiveRedisConnection 接口或 ReactiveRedisTemplate 类。这两个实体都提供了一个发布方法，该方法以需要发送的消息以及目标通道作为参数。ReactiveRedisConnection 接口需要原始数据（byte 类型数据）作为参数，而 ReactiveRedisTemplate 允许将任意对象作为消息传入：

```
//通过 ReactiveRedisConnection 发送消息
ByteBuffer msg = …
ByteBuffer channel = …
Mono<Long> publish = con.publish(msg, channel);

//通过 ReactiveRedisTemplate 发送消息
ReactiveRedisTemplate template = …
Mono<Long> publish = template.convertAndSend("channel", "message");
```

6.4.2 响应式消息订阅

在消息的接收端，可以通过直接命名或使用模式匹配订阅一个或多个频道。后一种方法非常有用，因为它不仅允许使用一个命令创建多个订阅，而且还可以侦听订阅时尚未创建的频道（只要它们与模式匹配）。

在底层，ReactiveRedisConnection 接口提供了 subscribe()和 pSubscribe()方法，分别对应按频道订阅的 subscribe 命令和按模式订阅的 psubscribe 命令。这两个命令的语法如下：

```
subscribe channel [channel ...]
psubscribe pattern [pattern ...]
```

其中，channel 表示频道名称，pattern 表示模式字符串。

可以使用多个通道或模式作为参数。要更改订阅，RedisConnection 接口提供了 getSubscription()方法和 isSubscribed()方法。Spring Data Redis 中的响应式订阅命令是非阻塞的。也就是说，在连接上调用 subscribe()方法会导致当前线程阻塞，等待订阅的消息。只有当订阅被取消时，线程才会被释放。可以在另一个线程上对订阅消息的连接调用 unsubscribe()或 pUnsubscribe()方法释放。

如上所述，一旦订阅，连接就开始等待消息。除了添加新订阅或修改、取消现有订阅外，不能在连接上使用其他命令。即只能使用 SUBSCRIBE、PSUBSCRIBE、UNSUBSCRIBE 或 PUNSUBSCRIBE 命令。

6.4.3 消息侦听器容器

由于其阻塞特性，低级订阅不具吸引力，因为它需要对每个侦听器进行连接和线程管理。为了解决这个问题，Spring Data 提供了 ReactiveRedisMessageListenerContainer 类

(org.springframework.data.redis.listener.RedisMessageListenerContainer),它可以完成转换和订阅状态管理任务。

ReactiveRedisMessageListenerContainer类充当消息侦听器容器。它用于从Redis频道接收消息,并公开一个消息流,该消息流发出已被反序列化的频道消息。消息侦听器容器负责注册以接收消息、获取和释放资源、异常转换等。这允许应用程序开发人员编写与接收消息相关的业务逻辑,并将基础的Redis问题委托给框架。消息流在发布者被订阅时在Redis中注册,如果订阅被取消,则取消注册。

6.5 小结

本章简要介绍了Redis的响应式操作接口。响应式编程是一种使用异步数据流的编程范式。可以将传统的命令式编程和响应式编程做个比较。以快餐店为例,在传统的命令式编程中,顾客到柜台点餐,服务员会将订单传给厨师,在食物准备好前会等待几分钟,再将食物送给顾客。然后,服务员开始为下一个顾客服务。这种方式极大浪费了时间,效率低下。首先,服务员等待厨师制作食物的时间被白白浪费掉了,在此期间除了等待没有做任何事情。而在响应式编程中,服务员从顾客手里接收订单,并传给厨师。此时,服务员无须等待厨师准备好食物,而是告诉顾客,一旦订单完成会通知他们。此时服务员可以继续为下一个顾客服务。这样,服务员可以一直接待顾客,直到厨师通知服务员某一笔订单已完成。当某订单完成时,服务员可以将食物递给顾客,之后可以继续接收订单。这样可以极大提高效率。首先,服务员没有空闲时间;其次,厨师一直忙于完成服务员送来的订单,厨师资源也被充分利用;最后,用户点餐也无须漫长等待,提升了用户的满意度。上述例子中,顾客相当于用户输入,厨师相当于CPU资源,服务员相当于一个服务线程。在响应式编程中,当CPU要处理一些比较耗时的操作(如数据库读写、文件读写、网络传输、文件打印等)时,用户无须等待操作执行结束,而是以订阅的方式等待操作结果,同时用户可以执行其他操作(下达其他指令)。当操作结束时,相关的操作线程会以流的形式发送操作结果,由于客户端订阅了相关的流,可以通过流接收操作结果。同时,响应式编程还提供了背压机制,用于向发布者提供反馈,使流中数据的发送速度与订阅者的处理速度匹配,实现了对计算资源的过载保护功能。这样,不但提高了CPU的利用率,还提升了用户体验。

对于以响应方式操作Redis,本章介绍了操作Redis中常用的数据类型(字符串、列表、哈希、集合、有序集合、地理空间数据)的方法。最后,简要介绍了响应式发布-订阅。目前,Spring支持的响应式编程框架是Reactor。在Reactor框架中,经常用函数式编程来处理响应流。因此,要学好响应式编程,还需要有函数式编程的基础。

第 7 章

视频讲解

Redis集群

前面章节中介绍的 Redis 都是在一台服务器上工作的。在项目开发中，一般不会简单地只在一台服务器上部署 Redis，因为单台 Redis 服务器不能满足高并发的要求。另外，如果该 Redis 服务器失效，整个系统就可能崩溃。为提升 Redis 服务器的吞吐量和可用性，通常会采用 Redis 集群。Redis 集群有三种模式：主从复制模式、哨兵模式和集群模式。一般会用主从复制集群实现数据热备份，用哨兵模式集群来保证可用性，用(分片)集群来提升吞吐量和可用性。本章主要介绍主从复制集群、哨兵模式集群和 Redis 分片集群。

7.1 主从复制集群

7.1.1 主从复制集群概述

主从复制集群也称为主从模式，主从复制集群中的 Redis 服务器可划分为主节点(Master)和从节点(Slave)。默认情况下，Redis 服务器都是主节点。一个主节点可以同时拥有多个从节点，而每个从节点只能对应一个主节点。当应用程序向主节点写入数据时，主节点通过 Redis 同步机制将数据复制到从节点上。这样一旦主节点出现故障，应用程序可以切换到从节点去读取数据，从而提升系统的可用性。配置主从复制集群后，主节点负责读写服务，从节点负责只读服务。主从复制模式里默认的读写分离机制能够提升系统缓存的读写性能。复制的数据流是单向的，只能从主节点复制到从节点，对于从节点的任何修改主节点都无法感知。修改从节点数据会造成主从节点数据不一致，因此建议不要修改从节点默认的只读模式。Redis 在主节点一端是非阻塞的。也就是说，对于 2.8 及以后的版本，Redis 支持采用异步的复制模式，即在进行主从复制时不影响主节点上的数据读写操作。

Redis 的主从复制拓扑结构可以支持单层或多层复制关系。根据拓扑复杂性可以分为以下三种：一主一从、一主多从、树状主从结构。

主节点 —复制的数据流→ 从节点

图 7-1 一主一从结构

一主一从结构是最简单的主从复制拓扑结构，如图 7-1 所示。一主一从结构可用于以下场景：主节点出现宕机时由从节点提供故障转移支持。

一主多从结构也称为星状拓扑结构，使得应用端可以利用多个节点实现读写分离，如图 7-2 所示。对于读操作占比较大的场景，可以把读命令发送到从节点来减轻主节点的压力。同时，在开发中如果需要执行一些比较耗时的读命令，如 KEYS、SORT 等，可以在其中一个节点上执行，防止耗时较长的查询对主节点造成阻塞从而影响缓存服务的稳定性。对于写并发量较高的场景，多个从节点会导致主节点写命令的多次发送从而过度消耗网络带宽，同时也增加了主节点的负载，影响服务的稳定性。

树状主从结构使得从节点（如图7-3中的从节点1和从节点2）不但可以复制主节点的数据，同时可以作为其他节点的主节点继续向下层复制。数据写入主节点后会同步到从节点1和从节点2，从节点1再把数据同步到从节点3和从节点4，从节点2把数据同步到从节点5。这样实现了数据一层一层向下复制。通过引入复制中间层，可以有效降低主节点负载和需要传送给从节点的数据量。当主节点需要挂载多个从节点时，为了避免对主节点的性能干扰，可以采用树状主从结构降低主节点的压力。

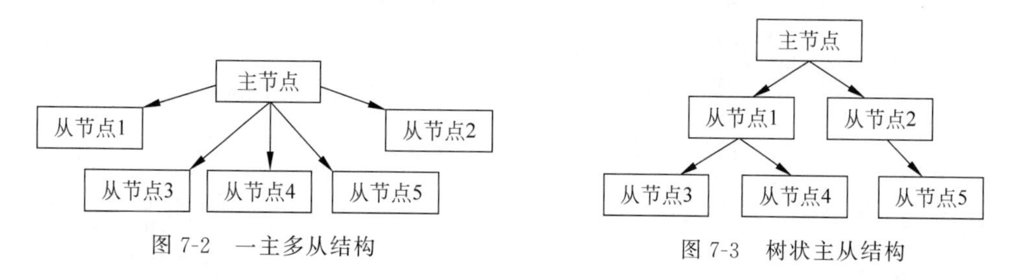

图7-2 一主多从结构

图7-3 树状主从结构

7.1.2 搭建主从复制集群

本节以搭建一个一主二从结构的主从复制集群为例，介绍搭建主从复制集群的两种方式（第三种方式在7.1.3节中介绍）。

1. 通过配置搭建主从复制集群

具体步骤如下。

第一步，启动主服务器。在命令行执行redis-server命令开启Redis主服务器，并且仍然使用默认的6379端口。

第二步，配置从服务器。指定两个从服务器分别监听8001和8002端口，分别创建两个文件夹，名为8001和8002。将主服务器目录中的redis.conf文件复制到上述两个文件夹中。修改8001文件夹中的redis.conf文件中的两行（在redis.conf文件中这两行并不相邻）：

```
1    port 8001
2    replicaof 127.0.0.1 6379
```

其中，第1行设定当前Redis服务器的端口号为8001，第2行配置该Redis服务器为从服务器，并连接到端口为6379的主服务器上。8002文件夹中的配置文件可做类似修改。

第三步，启动主从服务器。主服务器已经作为系统服务启动。在新的命令行窗口里运行redis-server{REDIS_HOME}/8001/redis.conf命令，开启从服务器。注意，此处的"{REDIS_HOME}/8001/"是一个相对路径，代表从服务器的配置文件的存放路径。端口为8001的从服务器启动后，启动界面如图7-4所示。可以用类似的方法启动端口为8002的从服务器。

第四步，开启从服务器的客户端。可以在新的命令行窗口执行命令：redis-cli-p 8001，开启端口为8001的从服务器的客户端。同时可以通过info replication命令查看主从复制集群的配置效果。info replication命令的部分执行结果如下：

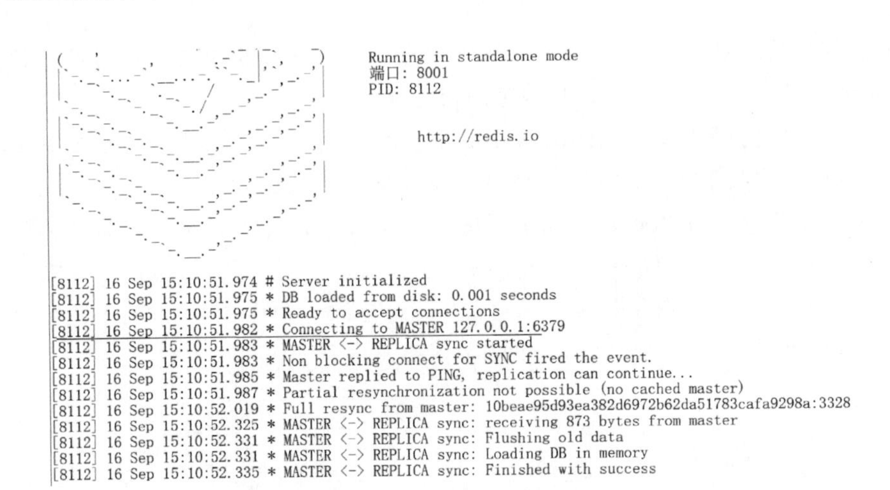

图 7-4 端口为 8001 的从服务器的启动界面

```
1   127.0.0.1:8001> info replication
2   # Replication
3   role:slave
4   master_host:127.0.0.1
5   master_port:6379
```

类似地，可以开启端口为 8002 的从服务器的客户端。至此，完成了通过配置文件搭建主从复制集群的任务。

2. 通过命令搭建主从复制集群

主从复制集群除了采用配置参数的方式搭建外，还可以采用命令方式搭建。具体步骤如下。

第一步，与“通过配置搭建主从复制集群”方式中的第一步完全一致。

第二步，编写从服务器的配置文件。复制主服务器的配置文件 redis.conf 到指定文件夹下，如/usr/local/etc/redis-cluster/9000 文件夹，并修改从服务器的端口。

```
port 9000
```

第三步，在启动从服务器的命令 redis-server 后加入命令参数“--slave of{masterHost}{masterPort}”开启从服务器。如执行命令：redis-server/usr/local/etc/redis-cluster/9000/redis.conf--slave of 127.0.0.1 6379。该命令读取 9000 文件夹下的 Redis 配置文件 redis.conf，并将端口为 9000 的 Redis 服务器作为端口为 6379 的 Redis 服务器的从机。采用同样的设置方法，可以将端口为 9001 的 Redis 服务器作为端口为 6379 的 Redis 服务器的从机，进而搭建起一主二从的主从复制集群。

第四步，使用“1. 通过配置搭建主从复制集群”方式中的第四步中的方法开启从客户端，并在从客户端利用 info replication 命令检查配置是否生效。

7.1.3 检验读写分离效果

在搭建了主从复制集群后，可以在从服务器的客户端执行 info replication 命令，在命令输出内容中可看到 slave_read_only:1 这项配置，说明从服务器默认是只读的。同时，可在从服务器的客户端执行命令 set hello redis，向从服务器写入数据，会看到下面的错误提示：

```
127.0.0.1:9000 > set hello redis
(error) READONLY You can't write against a read only replica.
```

为检验搭建好的主从复制集群的读写分离效果，可利用 RedisTemplate 类编写代码进行测试。例 7-1 针对 7.1.2 节中“2. 通过命令搭建主从复制集群”方式测试读写分离效果。

【例 7-1】 检验主从复制集群的读写分离效果。

具体步骤如下。

（1）创建配置文件。

在 src/main/resources 目录下编写一个名为 replication.properties 的配置文件，用于指定主从服务器的 IP 地址和端口。代码如文件 7-1 所示。

【文件 7-1】 replication.properties

```
1    master.ip = 127.0.0.1
2    master.port = 6379
3    …#此处略去了其他从服务器的 IP 地址和端口
4    slave3.ip = 127.0.0.1
5    slave3.port = 9000
```

（2）编写配置类。

配置类用于读取配置文件中设置的参数并实例化 RedisTemplate。代码如文件 7-2 所示。

【文件 7-2】 RedisTemplateConfig.java

```
1    @Configuration
2    @PropertySource("classpath:replication.properties")
3    @ComponentScan(basePackages = "com.example.redis.cluster.replication")
4    public class RedisTemplateConfig {
5       @Value("${master.ip}")
6       private String masterIp;
7       @Value("${master.port}")
8       private int masterPort;
9
10      @Value("${slave3.ip}")
11      private String slaveIp;
12      @Value("${slave3.port}")
13      private int slavePort;
14
15      @Bean(name = "master")
16      public RedisConnectionFactory redisConnectionFactory1() {
17         RedisStandaloneConfiguration rsc =
18              new RedisStandaloneConfiguration();
19         rsc.setHostName(masterIp);
20         rsc.setDatabase(0);
21         rsc.setPort(masterPort);
22         return new JedisConnectionFactory(rsc);
23      }
24      //创建名为 slave 的 RedisConnectionFactory 对象，代码此处略
25
26      @Bean(name = "mTemplate")
```

```
27        public StringRedisTemplate redisTemplate1(
28                @Autowired RedisConnectionFactory master){
29            StringRedisTemplate template = new StringRedisTemplate();
30            template.setConnectionFactory(master);
31            return template;
32        }
33        //创建名为 sTemplate 的 StringRedisTemplate 实例,用于操作从服务器,代码略
34    }
```

如文件 7-2 所示，第 1 行采用@Configuration 注解标注类，指示这个类声明一个或多个@Bean 方法，Spring 容器可以处理该类，以便在运行时创建这些 Bean。第 2 行采用@PropertySource 注解指定要读取的配置文件。在读取了配置文件后，用@Value 注解将配置文件中设置的属性值赋予该类的属性，如 masterIP（第 5、6 行）。第 15 行用@Bean 注解声明 redisConnectionFactory1()方法会创建一个名为 master 的 Bean。用类似的方法可创建一个名为 slave 的 RedisConnectionFactory 对象（Bean）。随后，第 27 行的 redisTemplate1()方法创建一个用于操作主服务器的 StringRedisTemplate 对象 mTemplate。可以用类似的方法创建用于操作从服务器的 StringRedisTemplate 对象。

（3）编写主从服务线程类。

在 src/main/java 目录下创建一个主服务线程类 MasterDemo，用来模拟对主服务器的写操作。代码如文件 7-3 所示。

【文件 7-3】 MasterDemo.java

```
1     package com.example.redis.cluster.replication;
2     //import 部分略
3     @Component(value = "masterDemo")
4     public class MasterDemo extends Thread{
5         @Resource(name = "mTemplate")
6         private StringRedisTemplate mTemplate;
7
8         public void run(){
9             mTemplate.opsForValue().set("hello","replication");
10        }
11    }
```

类似地，可以编写一个从服务线程类 SlaveDemo，用来模拟对从服务器的读操作，其中的 run()方法只需改为：

```
System.out.println(sTemplate.opsForValue().get("hello"));
```

（4）编写测试代码。

在 src/test/java 目录下创建一个测试类，用来检验主从复制集群的读写分离效果。测试代码如文件 7-4 所示。

【文件 7-4】 TestReplication3.java

```
1     package com.example.redis.replication;
2     @RunWith(SpringJUnit4ClassRunner.class)
```

```
 3    @ContextConfiguration(classes= RedisTemplateConfig.class)
 4    public class TestReplication3 {
 5
 6        @Resource
 7        private MasterDemo masterDemo;
 8        @Resource
 9        private SlaveDemo slaveDemo;
10
11        @Test
12        public void testReplication() {
13            try{
14                masterDemo.start();
15                Thread.sleep(1000);
16                slaveDemo.start();
17                Thread.sleep(200);
18            } catch(InterruptedException e) {
19                e.printStackTrace();
20            }
21        }
22    }
```

如文件 7-4 所示，在执行测试时，首先开启主节点写线程（主线程），向主节点以键 hello 写入值 replication（第 14 行）。完成写入操作后主线程暂停 1 秒，待主从复制集群完成数据同步后再执行读操作（第 15 行）。随后开启从节点读线程（从线程）（第 16 行）。延迟 0.2 秒，等待所有线程执行完毕，结束测试（第 17 行）。运行该测试代码，从线程在控制台的输出结果如图 7-5 所示。

```
✔ Tests passed: 1 of 1 test – 1 s 224 ms
C:\software\jdk-11.0.13\bin\java.exe ...
SLF4J: Failed to load class "org.slf4j.impl.StaticLoggerBinder".
SLF4J: Defaulting to no-operation (NOP) logger implementation
SLF4J: See http://www.slf4j.org/codes.html#StaticLoggerBinder for further details.
replication

Process finished with exit code 0
```

图 7-5　从线程在控制台的输出结果

此外，通过 RedisTemplate 类也可以设置主从复制集群并检验读写分离效果。

【例 7-2】 以编程方式搭建主从复制集群。

注意，例 7-2 中并没有预先设置 Redis 服务器间的主从复制关系，即初始时只需设置两台独立的 Redis 服务器，一台监听 6379 端口，准备作为主节点使用（以下称 6379 主机）；另一台监听 9000 端口，准备作为从节点使用（以下称 9000 主机）。

第一步完成配置文件的创建；第二步编写配置类。以上两步骤的做法与例 7-1 完全相同，此处不再赘述。第三步，编写测试代码，如文件 7-5 所示。

【文件 7-5】 TestReplication2.java

```
1    package com.example.redis.replication;
2    //import 部分略
3    @RunWith(SpringJUnit4ClassRunner.class)
4    @ContextConfiguration(classes= RedisTemplateConfig2.class)
```

```
5    public class TestReplication2 {
6        @Autowired
7        private StringRedisTemplate sTemplate;
8        @Autowired
9        private StringRedisTemplate mTemplate;
10
11       @Test
12       public void testReplication() {
13           mTemplate.opsForValue().set("hello","replication");
14           sTemplate.slaveOf("127.0.0.1",6379);
15           assertEquals("replication",
16                   sTemplate.opsForValue().get("hello"));
17           sTemplate.slaveOfNoOne();
18       }
19   }
```

如文件 7-5 所示，测试类首先绑定两个 StringRedisTemplate 对象：mTemplate 和 sTemplate。这两个对象分别用来操作 6379 主机和 9000 主机（第 6～9 行）。随后，先向主节点（6379 主机）以键 hello 写入值 replication（第 13 行），随后调用 RedisTemplate 类的 slaveOf()方法，将 sTemplate 所操作的 9000 主机设置为从节点（第 14 行）。即配置主从复制集群的第三种方法为：调用 RedisTemplate 类的 slaveOf()方法。第 15、16 行应用结果断言来判定从 9000 主机中读取的数据是否正确。最后，从节点（9000 主机）调用 slaveOfNone()方法解除与主节点（6379 主机）的复制关系（第 17 行）。

解除复制关系后从节点的变化体现在以下两点：①从节点断开与主节点的复制关系，但不抛弃原有数据，只是无法再获取主节点上的数据；②从节点晋升为主节点。

通过调用 slaveOf()方法还可以实现切主操作。切主操作是指把从节点对当前主节点的复制关系切换为复制另一个主节点。实现切主操作只要调用 RedisTemplate 类的 void slave(newMasterIP，newMasterPort)方法即可。切主操作的主要流程如下：

（1）中断与旧主节点的复制关系。

（2）与新主节点建立复制关系。

（3）删除从节点当前所有数据。

（4）对新主节点进行复制操作。

在主从复制集群中，数据集可以存在多个副本（从节点）。这些副本可以应用于读写分离、故障转移、实时备份等场景。例如，对于读占比较高的场景，可以通过把一部分读流量分摊到从节点来减轻主节点的压力，同时需要注意永远只对主节点进行写操作。由于复制只能从主节点到从节点，对于从节点数据的任何修改主节点都无法感知，修改从节点数据会造成主从节点数据不一致。因此，不建议修改从节点的只读模式。

7.2 哨兵模式集群

使用主从复制集群一方面可以提升 Redis 的可用性，如主节点失效后，可以启用从节点上的备份数据；另一方面也可以通过读写分离来提升性能。但是当主节点发生故障后，从节点无法自动承担主节点的任务，需要人工将从节点晋升为主节点，同时要通知应用程序更新主节点地址。这样会造成一段时间内服务器处于不可用状态,同时数据安全得不到保障。也就是说,主从复制集群在主节点发生故障时可能会出现数据丢失的情况。如果在主节点发生故障时,从节点能够自动变成主节点,就可以解决上述问题。这就是本节要介绍的哨兵模式集群。

7.2.1 哨兵模式集群概述

Redis 从 2.8 版本开始提供了稳定版的 Redis Sentinel(哨兵)机制。一般来说,哨兵机制会和主从复制模式整合使用,即组建哨兵模式集群。在哨兵模式集群中,由一台或多台服务器引入哨兵进程。引入哨兵进程的服务器称为哨兵节点。

哨兵节点一般不存储数据,它们的作用是监控 Redis 集群中主节点的工作状态。在主节点发生故障时,哨兵节点会主导故障转移流程,实现主节点和从节点的切换。具体来讲就是在该主节点的下属从节点中选出一个新的主节点,并完成相应的配置更改等任务,从而保证系统的高可用性。

哨兵的作用如下:

(1) 监控。哨兵节点会定期检测 Redis 数据节点和其他哨兵节点是否正常工作。

(2) 通知。哨兵节点会将故障转移的结果通知给应用程序。

(3) 自动故障恢复。如果主节点故障,哨兵进程会开始一次故障转移操作。它会将失效主节点的其中一个从节点晋升为新的主节点,并让失效的主节点的其他从节点改为复制新的主节点。当客户端试图连接失效的主节点时,哨兵节点也会向客户端发送新主节点的地址。

(4) 提供配置。在哨兵模式下,客户端在初始化时连接的是哨兵节点集合,从中获取主节点信息。

哨兵模式集群架构如图 7-6 所示。其中,哨兵节点集合包含若干哨兵节点,这样即使个别哨兵节点不可用,整个哨兵节点集合依然可以工作。哨兵模式集群的工作原理见 7.2.3 节。

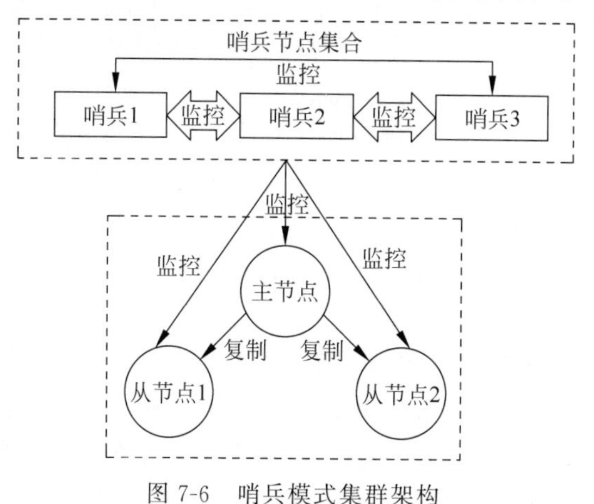

图 7-6 哨兵模式集群架构

哨兵模式集群中,对于节点的故障判断是由多个哨兵节点共同完成的,这样可以有效防止误判。在生产环境中,不宜将哨兵服务器部署在同一台物理计算机上,并且由于判定主节点客观下线需要半数以上的哨兵做出相同的下线判定,因此应该部署至少 3 个且奇数个哨兵节点。

7.2.2 搭建哨兵模式集群

搭建一个由3个哨兵节点、1个主节点、2个从节点组成的哨兵模式集群。条件所限，本书的哨兵模式集群是在一台计算机上创建的，所以集群中所有节点的IP地址都为127.0.0.1。集群拓扑结构如图7-6所示，物理部署方案如表7-1所示。

表7-1 哨兵模式集群的物理部署方案

角色	IP地址	端口	别名
主节点(master)	127.0.0.1	6379	6379节点
从节点(slave)1	127.0.0.1	8001	8001节点
从节点(slave)2	127.0.0.1	8002	8002节点
哨兵(sentinel)1	127.0.0.1	26379	26379节点
哨兵(sentinel)2	127.0.0.1	26380	26380节点
哨兵(sentinel)3	127.0.0.1	26381	26381节点

默认情况下，哨兵进程会监听26 379端口。本节在7.1.2节搭建的主从复制集群的基础上添加3个哨兵节点，组建哨兵模式集群。3个哨兵节点的部署方法是完全一致的(端口不同)，下面以26 379节点的部署为例进行说明，具体步骤如下。

1. 配置哨兵节点

创建一个名为26 379的文件夹，并将Redis安装路径下的redis-sentinel.conf文件复制到该文件夹下，并做出如下修改(注意在该文件中，以下内容并不是物理相邻的)：

```
1    port 26379
2    #每个长期运行的进程都需要一个工作目录.默认设置是/tmp 文件夹
3    logfile "26379.log"
4    #命令格式: sentinel monitor <mastername> <ip> <redis-port> <quorum>
5    #这里的 mymaster(名字)可以自定义
6    Sentinel monitor mymaster 127.0.0.1 6379 2
```

2. 启动哨兵节点

可以用redis-sentinel命令或redis-server命令启动哨兵节点，两种方式本质上是一致的，具体命令如下。

方法一：redis-sentinel redis-sentinel.conf。

方法二：redis-server redis-sentinel.conf-sentinel。

3. 确认

哨兵节点本质上是一个特殊的Redis节点，所以也可以通过info命令来查询它的相关信息(前提是3个哨兵节点均已配置完毕)。从下面的info命令的返回结果片段来看，哨兵节点找到了主节点127.0.0.1:6379，发现了它的两个从节点，同时发现一共有3个哨兵节点：

```
#执行命令
$ redis-cli -h 127.0.0.1 -p 26379 info sentinel
#执行命令后返回的结果片段
#Sentinel
sentinel_masters:1
sentinel_tilt:0
```

```
sentinel_tilt_since_seconds: - 1
sentinel_running_scripts:0
sentinel_scripts_queue_length:0
sentinel_simulate_failure_flags:0
master0:name = mymaster, status = ok, address = 127.0.0.1:6379, slaves = 2, sentinels = 3
```

也可以通过某哨兵节点查看主节点信息，执行下述命令开启 26380 节点的客户端：

```
$ redis - cli  - p 26380
```

随后在 26380 节点上执行 sentinel master mymaster 命令查看主节点信息，执行该命令后的输出内容片段如下：

```
127.0.0.1:26380 > sentinel master mymaster
 1) "name"
 2) "mymaster"
 3) "ip"
 4) "127.0.0.1"
 5) "port"
 6) "6379"
 7) "runid"
 8) "59264441b301f5a0fc68fb2776f0aa4181584c94"
 9) "flags"
 10) "master"
```

至此，哨兵模式集群搭建完毕。通过哨兵模式的配置可以看出哨兵模式集群是基于主从复制集群的。哨兵模式集群具有主从复制集群的所有优点，是主从复制集群的升级，能够实现自动化的故障恢复。

7.2.3 哨兵节点的常用配置

Redis 安装目录下有一个 redis-sentinel.conf 文件，这是默认的哨兵节点配置文件。现将其部分内容摘录如下。

```
1    port 26379
2    dir /opt/soft/redis/data
3    sentinel monitor mymaster 127.0.0.1 6379 2
4    sentinel down - after - milliseconds mymaster 3000
5    sentinel parallel - syncs mymaster 1
6    sentinel failover - timeout mymaster 180000
```

其中，port 和 dir 分别代表哨兵节点的端口和工作目录。第 3 行用于配置哨兵节点所监控的主节点，命令格式为：

```
sentinel monitor <master - name> <ip> <redis - port> <quorum>
```

上述配置文件的片段说明哨兵节点要监控的是一个名为 mymaster、IP 地址为 127.0.0.1、端口为 6379 的主节点。其中，参数 master-name 是自定义的，它只能包含英文字母、数字和“-”(英文减号)字符；参数 quorum 代表要判定主节点客观下线所需要的票数。哨兵节点基于心跳机制监测节点的服务状态。每个哨兵节点每隔 1 秒向集群中的所有节点(主、从节点和其他哨兵节点)发送 PING 命令做一次心跳检测，确认当前这些节点是否可达。如果某哨兵节点发现某节点未在规定时间做出响应，则认为该节点主观下线。当哨兵节点发现主节点主观下线时，该哨兵节点会通过命令向其他哨兵节点询问对主节点的判断，若超过指定数量（数量由 quorum 参数指定）的哨兵都认为该主节点主观下线，则该主节点客观下线。因此，建议将参数 quorum 的值设置为哨兵节点数量的一半加 1。哨兵节点会对所有节点进行监控。在哨兵节点的配置中没有看到有关从节点和其余哨兵节点的配置信息，是因为哨兵节点会从主节点中获取有关从节点和其余哨兵节点的相关信息。

每个哨兵节点都要通过定期发送 PING 命令来判断 Redis 数据节点和其余哨兵节点是否可达，如果超过了 down-after-milliseconds 配置的时间（第 4 行）且没有有效回复，则判定该节点不可达。命令格式为：

```
sentinel down-after-milliseconds <master-name> <time>
```

其中，参数 time 就是超时时间，单位：毫秒。这个配置项是判定节点是否可达的重要依据。

当哨兵节点集合对主节点主观下线的判定达成一致时，所有的哨兵节点会进行领导选举工作。选择一个哨兵节点作为领导者负责故障转移。哨兵领导者会从从节点中选择一个作为新的主节点。其余的从节点会向新的主节点发起复制操作。并且将客观下线的主节点更新为从节点，当其恢复后命令它去复制新的主节点。配置项 parallel-syncs 就是用来限制在一次故障转移之后，向新的主节点发起复制操作的从节点的个数（第 5 行）。命令格式为：

```
sentinel parallel-syncs <master-name> <num>
```

如果 num 的值配置得比较大，那么多个从节点会向新的主节点同时发起复制操作，尽管复制操作通常不会阻塞主节点，但是多个从节点同时向主节点发起复制，必然会给主节点所在机器带来一定的网络和磁盘读写开销。parallel-syncs 参数设置的效果如图 7-7 所示。其中，当 num=3 时从节点发起并行复制，num=1 时会发起顺序复制。

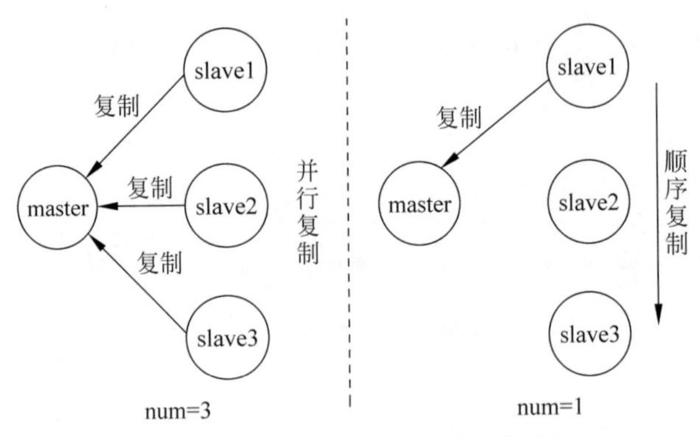

图 7-7 parallel-syncs 参数设置的效果

第 6 行的设置项 failover-timeout 通常被解释为故障转移的超时时间，它作用于故障转移的各个阶段：①选出合适的从节点。②将选出的从节点晋升为主节点。③命令其余从节点复制新的主节点。④等待原主节点恢复后命令它去复制新的主节点。命令格式为：

```
sentinel failover-timeout <master-name> <time>
```

其中，参数 time 的单位为毫秒。如第 6 行所示，如果在 180 秒内还没有完成主从服务器的切换，则判定本次恢复动作失败。

7.2.4 检验自动恢复效果

本节在7.2.2节搭建的哨兵模式集群基础上，使用 Spring Data Redis API 编写程序，检验主服务器失效后故障自动恢复的效果，步骤如下。

1. 创建配置类

编写一个名为 RedisSentinelConfig 的配置类，用于配置应用程序与哨兵节点的连接，代码如文件7-6所示。

【文件 7-6】 RedisSentinelConfig.java

```
1   @Configuration
2   public class RedisSentinelConfig {
3       @Bean
4       public RedisConnectionFactory jedisConnectionFactory() {
5           RedisSentinelConfiguration sentinelConfiguration =
6                   new RedisSentinelConfiguration().master("mymaster")
7                           .sentinel("127.0.0.1",26379)
8                           .sentinel("127.0.0.1",26380)
9                           .sentinel("127.0.0.1",26381);
10          return new JedisConnectionFactory(sentinelConfiguration);
11      }
12      @Bean
13      public RedisSentinelConnection redisConnection(
14              @Autowired RedisConnectionFactory rcf){
15          return rcf.getSentinelConnection();
16      }
17  }
```

如文件7-6所示，第3～11行创建 Redis 连接工厂。此处利用哨兵节点的配置创建与哨兵节点集合的连接。如果要创建与 Redis 数据节点的连接，可参考文件7-2。其中第6行指定当前哨兵模式集群中的主节点，只要用哨兵节点配置文件 redis-sentinel.conf 中配置的主节点的名字即可(内容见7.2.3节的配置项 sentinel monitor)。随后，第7～9行指定3个哨兵节点的地址和端口，并利用上述配置创建 Redis 连接工厂(第10行)。第12～16行通过已创建的连接工厂获得与哨兵节点集合的连接。

2. 编写测试类

在创建了与哨兵节点集合的连接后，可以调用 RedisSentinelCommands 接口(org.springframework.data.redis.connection.RedisSentinelCommands，RedisSentinelConnection 接口的父接口)的 masters()方法和 slaves()方法获取集群中主节点和从节点的对象(org.springframework.data.redis.connection.RedisServer)集合。代码如文件7-7所示。

【文件 7-7】 TestSentinel.java

```
1   @RunWith(SpringJUnit4ClassRunner.class)
2   @ContextConfiguration(classes = RedisSentinelConfig.class)
3   public class TestSentinel {
4       @Autowired
5       private RedisSentinelConnection rsc;
6
7       @Test
```

```
8       public void testSentinel(){
9          Collection<RedisServer> master_collection = rsc.masters();
10         System.out.println("master list :");
11         Iterator master_iterator = master_collection.iterator();
12         RedisServer master = null;
13         if(master_iterator.hasNext()){
14            master = (RedisServer)master_iterator.next();
15            System.out.println(master.getPort());
16         }
17         System.out.println("slave list : ");
18         Collection<RedisServer> slave_collection = rsc.slaves(master);
19         Iterator<RedisServer> slave_iterator =
20              slave_collection.stream().iterator();
21         slave_iterator.forEachRemaining(e ->
22              System.out.println(e.getPort()));
23      }
24   }
```

如文件 7-7 所示，第 9 行调用 RedisSentinelCommands 接口的 masters()方法获取哨兵节点集合所监控的所有主节点对象及其状态。该方法的原型为：

```
Collection<RedisServer> masters()
```

目前的集群部署如表 7-1 所示，集群中只有一个主节点。因此，第 13～16 行使用 if 语句输出迭代器中的内容。此处调用 RedisNode 类（org. springframework. data. redis. connection. RedisNode，RedisServer 类的父类）的 getPort()方法输出了主节点的端口。在指定了主节点对象后，调用 RedisSentinelCommands 接口（org. springframework. data. redis. connection. RedisSentinelCommands）的 slaves()方法获取复制该主节点的从节点对象，其中 slaves()方法的原型为：

```
Collection<RedisServer> slaves(NamedNode master)
```

在运行测试代码前，要保证哨兵模式的所有节点均已开启且能够正常工作。在当前哨兵模式集群正常工作的情况下，运行该测试代码后，运行结果如图 7-8 所示。

随后，可以杀死 6379 线程，模拟 6379 主机出现故障。再次运行测试代码，查看集群的故障自动恢复效果，如图 7-9 所示。

图 7-8　哨兵模式集群正常工作的情况

```
master list :
8001
slave list :
8002
6379
```

图 7-9　故障自动恢复效果

由图 7-9 可知，6379 主机因为故障客观下线，8001 主机成为新的主节点。6379 主机被设置为 8001 主机的从节点。

7.3　Redis 分片集群

主从复制集群实现了数据的热备份，哨兵模式集群实现了 Redis 服务的高可用。但是有两个问题这两种模式都没有解决：第一，这两种模式都只能有一个主节点负责写操作，在高并发的写操作场景，主节点就会成为性能瓶颈；第二，两种模式中从节点默认负责备份数据。这样造成了大量的数据冗余和资源浪费，而且内存扩展困难。

Redis 的集群模式(Redis Cluster)可以实现多个节点同时写操作，而且无须额外的监控节点。Redis 集群模式采用无中心结构，每个节点都保存数据，节点之间互相连接从而知道整个集群的状态，具备故障自我恢复的能力。鉴于 Redis Cluster 以分片形式存储数据，本书将 Redis Cluster 称为 Redis 分片集群，简称分片集群。

7.3.1 Redis 分片集群概述

Redis 分片集群是 Redis 的分布式实现，在 3.0 版本正式推出。Redis 分片集群是一个由多个主从节点组成的分布式服务器集群，它具有实时备份、高可用和高吞吐量的特性。分片集群会引入主从复制模式，从而实现数据的实时备份，当主节点宕机时会自动启用从节点。分片集群也内置了高可用机制，支持 n 个主节点，每个主节点都可以挂载多个从节点，当主节点宕机时，集群会提升它的某个从节点作为新的主节点。分片集群可以使数据自动分片到多个 Redis 节点从而提高吞吐量。Redis 分片集群架构如图 7-10 所示。

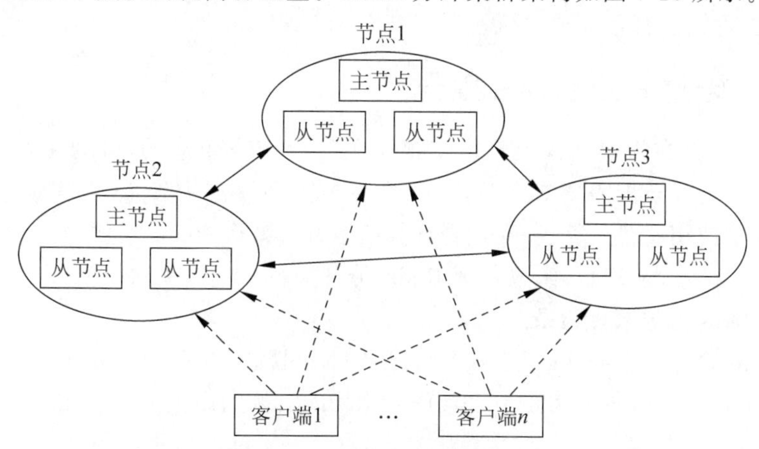

图 7-10 Redis 分片集群架构

如图 7-10 所示，Redis 分片集群采用去中心化的思想，没有中心节点，便于水平扩展。对于客户端来说，分片集群可以看成一个整体，可以连接任意一个节点进行操作，就像操作单一 Redis 实例一样，不需要任何中间件。当客户端操作的键没有分配到该节点上时，Redis 会返回转向指令，指向正确的节点。

对于分片集群来说，首先要解决的是如何把整个数据集按照分区规则映射到多个节点上，即把数据集划分到多个节点上，每个节点负责整体数据集的一个子集，如图 7-11 所示。常见的分区规则有哈希分区和顺序分区。

分片集群采用的是哈希分区规则中的虚拟槽分区算法。虚拟槽分区算法使用分散度良好的哈希函数把所有数据映射到一个固定范围的整数集合中，整数定义为槽(Slot)。这个范围一般远远大于节点数。如分片集群将所有数据存储区划分为 16 384 个槽，槽的范围为 0～16 383。每个节点负责一部分槽。槽的信息存储于每个节点中。如果集群中有 3 个主节点，分片集群会把 0～16 383 范围的槽分成三部分：如 A 节点负责的槽的范围为 0～5500，B 节点负责的槽的范围为 5501～11 000，C 节点则负责范围为 11 001～16 383 的槽。对于给

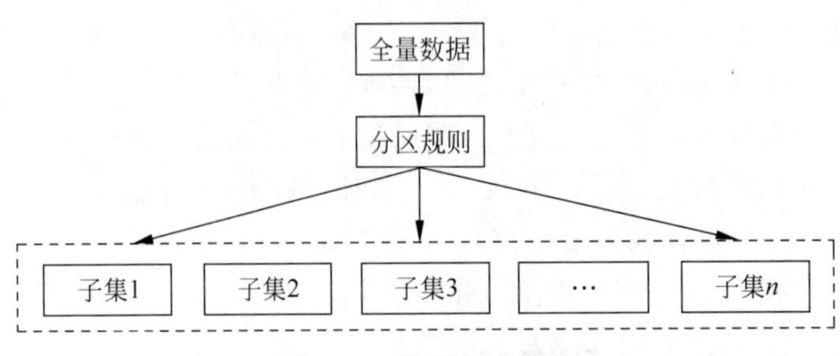

图 7-11 数据存储分区

定的键，Redis 会首先对键进行 CRC16 校验并对 16 384 取模，计算出键所在的槽，然后再到对应的槽上进行数据存取操作。槽的计算公式为 HashSlot=CRC16(key)mod 16 384。

分片集群的设计方案使得在集群中添加和删除节点变得很容易。例如，如果想添加一个新的节点 D，则需要将一些槽从节点 A、B、C 移动到 D。同样，如果想要从集群中删除节点 A，只需将 A 提供的槽移动到 B 和 C。一旦节点 A 为空，就可以将其从集群中完全删除。将槽从一个节点移动到另一个节点不需要停止任何操作，因此，添加和删除节点，或更改节点持有的槽的百分比，都不需要停机。

7.3.2 搭建 Redis 分片集群

Redis 分片集群在物理结构上是由集群上的多个节点组成的，分为主节点和从节点。一个 Redis 分片集群至少需要 6 个节点：3 个主节点和 3 个从节点(通常简称 3 主 3 从)。本节的集群环境使用一台计算机，在这台计算机上开启 6 个 Redis 实例，每个实例占用一个端口，用来模拟 3 主 3 从环境，组成一个 Redis 分片集群。具体步骤如下。

1. 创建 Redis 虚拟节点目录

使用 mkdir 命令在/usr/local/etc 下创建 cluster 目录，并在 cluster 目录下创建 6 个子目录：7000，7001，…，7005。这 6 个目录作为 Redis 实例的虚拟节点目录，分别用来存放 6 个 Redis 实例的配置文件。

2. 修改配置文件

使用 cp 命令将 Redis 的默认配置文件 redis. conf 复制到 7000 目录中，并命名为 redis-7000. conf。修改 redis-7000. conf 配置文件中的以下设置项(这些设置项在配置文件中并不相邻)：

```
1    #设置端口
2    port 7000
3    #实例以集群模式运行
4    cluster-enabled yes
5    #设置从节点为只读
6    replica-read-only yes
7    #指定集群配置文件
8    cluster-config-file nodes-7000.conf
```

其中，集群配置文件默认设置为 nodes-6379. conf。如果在同一台计算机上模拟 Redis 分片集群，会造成多个 Redis 实例读取同一个集群配置文件 nodes-6379. conf，这样会报告错误。因此，本节指定不同的 Redis 实例读取不同的集群配置文件。其余的 7001～7005 文件夹的设置方法与 7000 文件夹的设置方法完全相同。

3. 启动服务器

利用 cd 命令进入 7000 目录：$ cd cluster/7000 <回车>

在 7000 目录下执行下述命令：$ redis-server redis-7000. conf <回车>开启监听端口为 7000 的 Redis 实例，服务器开启后的界面如图 7-12 所示。可以同样的方法开启其余 5 个 Redis 实例。

```
7000 — redis-server 127.0.0.1:7000 [cluster] ptr_munge= — 80×24
691:M 31 Jan 2023 11:51:20.355 * No cluster configuration found, I'm 6a14d61c8e9
c33d42c17c407d8b94edd188f9eb7
```

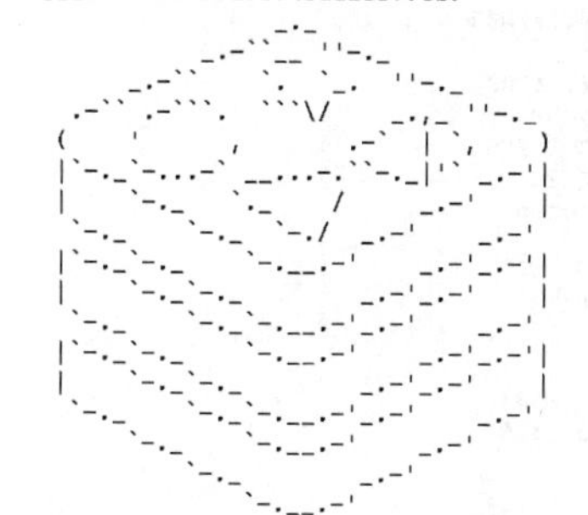

```
Redis 7.0.2 (00000000/0) 64 bit

Running in cluster mode
Port: 7000
PID: 691

https://redis.io

691:M 31 Jan 2023 11:51:20.375 # WARNING: The TCP backlog setting of 511 cannot
be enforced because kern.ipc.somaxconn is set to the lower value of 128.
691:M 31 Jan 2023 11:51:20.375 # Server initialized
691:M 31 Jan 2023 11:51:20.376 * Loading RDB produced by version 7.0.2
```

图 7-12　端口为 7000 的服务端开启后的界面

4. 创建集群

在默认的 Redis 的安装目录(如/usr/local/etc)中执行下述命令，创建集群。

```
$ cd /usr/local/etc
etc $ redis - cli -- cluster create 127.0.0.1:7000 127.0.0.1:7001
127.0.0.1:7002 127.0.0.1:7003 127.0.0.1:7004 127.0.0.1:7005 -- cluster - replicas 1
```

这里使用的命令是 CREATE，因为要创建一个新的集群。命令选项--cluster replica 1 意味着要为集群中的每个主节点创建一个副本。其他参数是要用于创建新集群的实例的地址列表。redis-cli 命令将提出一种配置，如图 7-13 所示。输入 yes 接受建议的配置。集群将按当前配置方案(如表 7-2 所示)进行创建。

表 7-2　redis-cli 提出的集群配置方案

主节点端口	从节点端口	槽
7000	7004	0～5460
7001	7005	5461～10 922
7002	7003	10 923～16 383

7.3.3 操作 Redis 分片集群

利用 Spring Data Redis API 操作分片集群通常有两种操作方式：第一是借助 RedisClusterConnection 接口；第二是借助 RedisTemplate 类。在操作 Redis 分片集群前，首先要搭建 Redis 分片集群(见 7. 3. 2 节)。随后，利用代码描述集群的基本情况。对于

```
xudemacbook-pro:~ xuguo$ redis-cli --cluster create 127.0.0.1:7000 127.0.0.1:7001 127.0.0.1:7002
127.0.0.1:7003 127.0.0.1:7004 127.0.0.1:7005 --cluster-replicas 1
>>> Performing hash slots allocation on 6 nodes...
Master[0] -> Slots 0 - 5460
Master[1] -> Slots 5461 - 10922
Master[2] -> Slots 10923 - 16383
Adding replica 127.0.0.1:7004 to 127.0.0.1:7000
Adding replica 127.0.0.1:7005 to 127.0.0.1:7001
Adding replica 127.0.0.1:7003 to 127.0.0.1:7002
>>> Trying to optimize slaves allocation for anti-affinity
[WARNING] Some slaves are in the same host as their master
M: 6a14d61c8e9c33d42c17c407d8b94edd188f9eb7 127.0.0.1:7000
   slots:[0-5460] (5461 slots) master
M: b7e2dd34722291c1825503263ac65ceac2e23d20 127.0.0.1:7001
   slots:[5461-10922] (5462 slots) master
M: 031c96c384f38da3f9ad740f7197db9564ce46a5 127.0.0.1:7002
   slots:[10923-16383] (5461 slots) master
S: 04f33c0ee344ee9c886bec22c5d3695967bc5487 127.0.0.1:7003
   replicates 031c96c384f38da3f9ad740f7197db9564ce46a5
S: 918566da8fd0d2d0448a33fa536c231d6531a479 127.0.0.1:7004
   replicates 6a14d61c8e9c33d42c17c407d8b94edd188f9eb7
S: e46032399c7219f932bba40fb0d42fc2e3ecfe6c 127.0.0.1:7005
   replicates b7e2dd34722291c1825503263ac65ceac2e23d20
Can I set the above configuration? (type 'yes' to accept): yes
>>> Nodes configuration updated
>>> Assign a different config epoch to each node
>>> Sending CLUSTER MEET messages to join the cluster
Waiting for the cluster to join
..
>>> Performing Cluster Check (using node 127.0.0.1:7000)
M: 6a14d61c8e9c33d42c17c407d8b94edd188f9eb7 127.0.0.1:7000
   slots:[0-5460] (5461 slots) master
   1 additional replica(s)
M: b7e2dd34722291c1825503263ac65ceac2e23d20 127.0.0.1:7001
   slots:[5461-10922] (5462 slots) master
   1 additional replica(s)
M: 031c96c384f38da3f9ad740f7197db9564ce46a5 127.0.0.1:7002
   slots:[10923-16383] (5461 slots) master
   1 additional replica(s)
S: 918566da8fd0d2d0448a33fa536c231d6531a479 127.0.0.1:7004
   slots: (0 slots) slave
   replicates 6a14d61c8e9c33d42c17c407d8b94edd188f9eb7
S: e46032399c7219f932bba40fb0d42fc2e3ecfe6c 127.0.0.1:7005
   slots: (0 slots) slave
   replicates b7e2dd34722291c1825503263ac65ceac2e23d20
S: 04f33c0ee344ee9c886bec22c5d3695967bc5487 127.0.0.1:7003
   slots: (0 slots) slave
   replicates 031c96c384f38da3f9ad740f7197db9564ce46a5
[OK] All nodes agree about slots configuration.
>>> Check for open slots...
>>> Check slots coverage...
[OK] All 16384 slots covered.
xudemacbook-pro:~ xuguo$
```

图 7-13　创建并配置分片集群

7.3.2 节的 3 主 3 从集群，可以创建名为 spring. redis. cluster. properties 的文件描述集群的一些属性，代码如文件 7-8 所示。

【文件 7-8】　spring. redis. cluster. properties

```
1    spring.redis.cluster.nodes =\
2      127.0.0.1:7000,\
3      127.0.0.1:7001,\
4      127.0.0.1:7002,\
5      127.0.0.1:7003,\
6      127.0.0.1:7004,\
7      127.0.0.1:7005
8    spring.redis.cluster.max-redirects: 3
```

其中，spring. redis. cluster. nodes 属性以逗号分隔的{主机}：{端口}的形式列举集群中的主机和端口(第 2～7 行)。spring. redis. cluster. max-redirects 属性则指定群集的重定向数。随后，读取集群属性文件，代码如文件 7-9 所示。

【文件 7-9】 MyClusterConfigurationProperties.java

```
1    package com.example.redis.cluster.cluster;
2    @Component
3    @PropertySource("spring.redis.cluster.properties")
4    public class MyClusterConfigurationProperties {
5        @Value("${spring.redis.cluster.nodes}")
6        private List<String> nodes;
7        //此处省略了 nodes 属性的 getters/setters 方法
8    }
```

1. 借助 RedisClusterConnection 接口操作 Redis 分片集群

应用程序与 Redis 集群或非集群的连接都是基于 RedisConnection 接口(org.springframework.data.redis.connection.RedisConnection)的。RedisClusterConnection 接口(org.springframework.data.redis.connection.RedisClusterConnection)是 RedisConnection 接口的子接口,它用来建立应用程序与 Redis 分片集群的连接,并将错误转换为 Spring DAO 层次异常。RedisClusterConnection 实例是由 RedisConnectionFactory 接口创建的,必须使用关联的 RedisClusterConfiguration 对象作为参数。文件 7-10 展示了如何利用 RedisClusterConnection 接口建立与分片集群的连接。

【文件 7-10】 MyRedisClusterConfig.java

```
1    @Configuration
2    @ComponentScan(basePackages = "com.example.redis.cluster.cluster")
3    public class MyRedisClusterConfig {
4
5        @Autowired
6        private MyClusterConfigurationProperties properties;
7        @Bean
8        public RedisConnectionFactory redisConnectionFactory(){
9            List<String> list = properties.getNodes();
10           return new JedisConnectionFactory(new
11               RedisClusterConfiguration(list));
12       }
13
14       @Bean
15       public RedisClusterConnection redisClusterConnection(@Autowired
16         RedisConnectionFactory rcf) {
17           return rcf.getClusterConnection();
18       }
19   }
```

如前所述,Redis 分片集群的行为与单一 Redis 实例甚至哨兵监控的主副本环境不同。这是因为自动分片机制会将一个键映射到 16 384 个槽中的某一个,这些槽分布在分片集群节点上。因此,涉及多个键的命令必须保证所有键映射到完全相同的槽,以避免跨槽错误。单个集群节点仅为一组专用的键提供服务。针对某个特定节点发出的命令仅返回该节点提供的键的结果。例如 KEYS 命令,当该命令发出到分片集群的节点时,它只返回请求发送到的节点提供的键,而不会返回分片集群中的所有键。因此,要获取分片集群环境中的所有的键，必须从所有已知主节点读取。

虽然特定键到相应槽节点的重定向由驱动程序库处理，但更高级的功能(如跨节点收集信息或向集群中的所有节点发送命令)可由 RedisClusterConnection 接口完成。文件 7-11 是一个从分片集群所有节点中提取键的例子。

【文件 7-11】 TestCluster.java

```
1   @RunWith(SpringJUnit4ClassRunner.class)
2   @ContextConfiguration(classes = MyRedisClusterConfig.class)
3   public class TestCluster {
4       @Autowired
5       private RedisClusterConnection connection;
6
7       @Test
8       public void testOpsForCluster(){
9           connection.set("thing1".getBytes(),"value1".getBytes());
10          connection.set("thing2".getBytes(),"value2".getBytes());
11          Set<byte[]> set = connection.keys("*".getBytes());
12          set.iterator().forEachRemaining(e -> System.out.println(
13            new String(e)));
14      }
15  }
```

如文件 7-11 所示，第 9、10 行调用 RedisClusterConnection 接口继承自 RedisStringCommands 接口的 set()方法，实现在集群中分别以 thing1 和 thing2 为键存储值 value1 和 value2。其中，RedisClusterConnection 接口继承了 RedisConnection、RedisStringCommands、RedisPubSubCommands 等众多 Redis 操作接口，调用这些接口中的方法，可以像操作普通的单一 Redis 实例一样操作 Redis 分片集群。第 11 行调用 RedisClusterConnection 接口继承自 RedisKeyCommands 接口的 keys()方法，获取集群中所有的键。其中的参数指定为"*"，相当于向集群中的每个主节点发送 KEYS * 命令，同时提取结果并返回键集合 set。对于表 7-2 中的分片集群配置，程序向端口为 7000、7001 和 7002 的主机发送 KEYS * 命令。如果仅请求单个节点的键，RedisClusterConnection 接口提供了重载的 keys()方法(例如，keys(RedisClusterNode node,byte[]pattern))。第 12、13 行则对返回的键集合执行遍历，并在控制台输出。

2. 借助 RedisTemplate 类操作 Redis 分片集群

RedisTemplate 类通过 ClusterOperations <K,V>子接口(org.springframework.data.redis.core.ClusterOperations<K,V>)提供对分片集群的访问。该接口的实例可以通过调用 RedisTemplate 类的 opsForCluster()方法获得。这样可以在集群内的单个节点上显式执行命令，同时保留为 RedisTemplate 配置的序列化和反序列化功能。ClusterOperations<K,V>子接口还提供管理命令，如可调用该接口的 meet()方法执行 CLUSTER MEET 命令，实现向集群中添加节点；或执行更高级的操作，如调用该接口的 reshard()方法对槽进行重新分配。文件 7-12 是一个利用 RedisTemplate 类从集群中端口为 7000 的主节点上获取所有键的操作案例。该案例代码是在文件 7-11 的基础上添加部分测试用例实现的。

【文件 7-12】 向文件 7-11 添加的部分测试代码(1)

```
1    @Autowired
2    private StringRedisTemplate template;
3    @Test
4    public void testOpsForCluster2(){
5      Set<String> keySet = template.opsForCluster()
6         .keys(new RedisClusterNode("127.0.0.1",7000),"*");
7      keySet.iterator().forEachRemaining(System.out::println);
8    }
```

其中,第 6 行调用了 ClusterOperations 子接口中的 keys()方法从指定集群节点获取所有键的集合。该方法的原型如下:

```
Set<K> keys(RedisClusterNode node,K pattern)
```

此外,还可以利用 ClusterOperations 子接口对象获取集群中全部节点的所有键的集合,代码如文件 7-13 所示。

【文件 7-13】 向文件 7-11 添加的部分测试代码(2)

```
1    @Test
2    public void testOpsForCluster3(){
3       ClusterOperations<String,String> ops = template.opsForCluster();
4       ValueOperations valueOperations = template.opsForValue();
5       valueOperations.set("key1","value1");
6       valueOperations.set("key2","value2");
7       valueOperations.set("key3","value3");
8       valueOperations.set("key4","value4");
9       for(int i = 7000;i <= 7005;i++){
10         RedisClusterNode node = new RedisClusterNode("127.0.0.1",i);
11         Set<String> keys = ops.keys(node,"*");
12         System.out.println(node + " ==> " +
13            Arrays.toString(keys.toArray()));
14      }
15   }
```

如文件 7-13 所示,第 4~8 行利用 ValueOperations 子接口向分片集群中写入数据,可以像操作单一 Redis 实例一样操作 Redis 分片集群。第 11 行调用 ClusterOperations 子接口的 keys()方法获取分片集群中所有节点的全部键集合(不建议在生产环境下使用,可能因为数据量过大造成阻塞)。运行此测试代码,程序运行结果如图 7-14 所示。从程序运行结果结合图 7-13 及表 7-2 的集群配置方案可知,端口为 7000 的主节点和端口为 7004 的从节点为相同槽的主从节点配置,它们存储的数据内容相同。同样,端口为 7001 的主节点和端口为 7005 的从节点存储的数据内容也是相同的。

```
127.0.0.1:7000 ==> [key3, key2]
127.0.0.1:7001 ==> [key1]
127.0.0.1:7002 ==> [key4]
127.0.0.1:7003 ==> [key4]
127.0.0.1:7004 ==> [key3, key2]
127.0.0.1:7005 ==> [key1]
```

图 7-14 文件 7-13 测试代码运行结果

从文件 7-12 和文件 7-13 的案例中可知,除了可以对 Redis 分片集群进行整体操作外,还可以通过 RedisClusterNode 对象指定分片集群中的某个节点,进而对分片集群中的某个节点单独发送命令。例如,可以调用 ClusterOperations 接口的 ping()方法向分片集群中的某节点发送 PING 命令,查看节点的状态。该方法的原型如下:

String ping(RedisClusterNode node)

案例代码如文件 7-14 所示。

【文件 7-14】 向文件 7-11 添加的部分测试代码(3)

```
1    @Test
2    public void testOpsForCluster4(){
3        ClusterOperations<String,`String> ops = template.opsForCluster();
4        for(int i = 7000;i<7006;i++){
5            RedisClusterNode node = new RedisClusterNode("127.0.0.1",i);
6            String ping = ops.ping(node);
7            System.out.println(node + " ==> " + ping);
8        }
9    }
```

```
127.0.0.1:7000 ==> PONG
127.0.0.1:7001 ==> PONG
127.0.0.1:7002 ==> PONG
127.0.0.1:7003 ==> PONG
127.0.0.1:7004 ==> PONG
127.0.0.1:7005 ==> PONG
```

图 7-15 文件 7-14 测试代码运行后的控制台输出

如文件 7-14 所示，第 4～6 行利用 ping()方法向集群中的节点逐一发送 PING 命令，即 ping()方法的返回值为节点的响应消息。通常情况下，集群中状态正常的节点在收到 PING 命令后会发送 PONG 响应消息。如果利用流水线或事务操作，则会返回 null。第 7 行则逐一输出集群中各个节点的返回消息。运行此测试代码后，控制台输出如图 7-15 所示。

对分片集群中的节点还可以进行数据库的相关操作，如手动刷新数据库、同步保存数据库快照、手动触发持久化。实现以上操作，可以分别调用 ClusterOperations 子接口的 flushDb()方法、save()方法和 bgSave()方法。这三个方法的原型如下：

(1) void flushDb(RedisClusterNode node)：刷新节点 node 上的数据库。

(2) void save(RedisClusterNode node)：同步保存节点 node 上当前数据库的快照。

(3) void bgSave(RedisClusterNode node)：在节点 node 上开启后台保存数据。

例如，对端口为 7001 的节点实行手动触发持久化，将该节点的数据保存到磁盘，代码如文件 7-15 所示。

【文件 7-15】 向文件 7-11 添加的部分测试代码(4)

```
1    @Test
2    public void testOpsForCluster5() {
3        ClusterOperations<String,String> ops = template.opsForCluster();
4        ops.bgSave(new RedisClusterNode("127.0.0.1",7001));
5    }
```

运行此测试代码，观察 7001 主机的控制台，输出信息如图 7-16 所示。

```
1290:M 20 Feb 2023 11:58:00.174 * Background saving started by pid 2018
2018:C 20 Feb 2023 11:58:00.179 * DB saved on disk
2018:C 20 Feb 2023 11:58:00.180 * Fork CoW for RDB: current 0 MB, peak 0 MB, ave
rage 0 MB
1290:M 20 Feb 2023 11:58:00.200 * Background saving terminated with success
```

图 7-16 7001 主机控制台输出

利用 ClusterOperations 子接口提供的方法还可以进行集群相关设置，如移动和添加槽、移除和添加节点、关闭节点、手动开启 AOF(Append-Only-File，只追加文件)持久化等。

相关方法的原型如下：

（1）void addSlots(RedisClusterNode node,int...slots)：将槽 slots 添加到节点 node。

（2）void addSlots(RedisClusterNode node,RedisClusterNode.SlotRange range)：将 range 范围中的槽添加到节点 node。

（3）void reshard(RedisClusterNode source,int slot,RedisClusterNode target)：将槽 slot 从一个源节点 source 移到目标节点 target,并复制与该槽关联的键。移动槽必须在主节点上面进行。

（4）void forget(RedisClusterNode node)：从集群中移除节点 node。为防止出现全量复制,建议先移除从节点,再移除主节点。

（5）void meet(RedisClusterNode node)：将节点 node 添加到集群中。

（6）Collection < RedisClusterNode > getSlaves(RedisClusterNode node)：获取节点 node 对应的所有从节点。

（7）void shutdown(RedisClusterNode node)：关闭节点 node。

（8）void bgReWriteAof(RedisClusterNode node)：在节点 node 上启动 AOF 重写过程。

例如,将当前集群中的主机全部关闭,代码如文件 7-16 所示。

【文件 7-16】 向文件 7-11 添加的部分测试代码(5)

```
1    @Test
2    public void testOpsForCluster6() {
3        ClusterOperations<String, String> ops = template.opsForCluster();
4        for(int i = 7000;i < 7006;i++){
5            RedisClusterNode node = new RedisClusterNode("127.0.0.1",i);
6            ops.shutdown(node);
7            System.out.println(node + " ==> shutdown.");
8        }
9    }
```

运行此测试代码后,控制台输出如图 7-17 所示。

```
127.0.0.1:7000 ==> shutdown.
127.0.0.1:7001 ==> shutdown.
127.0.0.1:7002 ==> shutdown.
127.0.0.1:7003 ==> shutdown.
127.0.0.1:7004 ==> shutdown.
127.0.0.1:7005 ==> shutdown.
```

图 7-17 文件 7-16 测试代码运行后的控制台输出

7.4 小结

本章介绍了 Redis 集群的 3 种模式：主从复制模式集群、哨兵模式集群和 Redis 分片集群。作为非关系数据库,Redis 本身就具有较高的读写性能,构建集群更主要的目的是提高 Redis 的可用性,防止单机 Redis 宕机时,缓存数据丢失或无法提供缓存服务的情况出现。而且,3 种集群模式中的从机默认为只读,可以仅做数据备份,从而提高可用性；也可以通过程序实现读写分离,从而提高性能。

对于主从复制模式集群,参与集群的 Redis 服务器被划分为主节点和从节点,每个从节点只能对应一个主节点,而一个主节点可以对应多个从节点。从节点可以从主节点复制数据,而且数据流是单向的。主从复制模式具有下列优点:①主节点可以自动将数据同步到从节点。若主节点宕机,可手动将从节点晋升为主节点,从而保证可用性。②主从节点之间的数据同步是以非阻塞方式进行的。在数据同步期间,客户端仍然可以查询或更新数据。同时,主从复制模式也有一些缺点:①主节点和从节点之间的数据同步是非阻塞的,在读写分离应用场景下,可能会导致复制数据延迟、读到过期数据等不一致问题。②从节点对主节点的数据备份一般是全量复制,会造成大量的数据冗余。③当一个主节点挂载多个从节点或一台机器上部署多个主节点的情况下有引起复制风暴的风险。即大量的从节点对主节点进行全量复制,可能造成主节点所在的机器网络带宽耗尽。④难以支持在线扩容。

为了解决主从复制模式中当主节点由于故障无法提供服务时,需要人工将从节点晋升为主节点的问题,从 2.8 版本开始,Redis 支持哨兵模式集群。

哨兵模式集群是一个分布式架构,其中包含若干哨兵节点和 Redis 数据节点。每个哨兵节点会对数据节点和其余哨兵节点进行监控。当哨兵节点发现某节点不可达时,则标记该节点下线。如果发现主节点不可达时,哨兵节点判定主节点为主观下线,当有超过半数的哨兵节点做出同样的结论时,则判定主节点为客观下线。当主节点被标记为客观下线时,哨兵集群会选举一个领头哨兵自动将从属于客观下线的主机的某从节点晋升为主节点,完成故障转移。故障的判定和恢复过程可以完全自动化,并且这一过程需要哨兵投票。因此,一般需要奇数个(至少 3 个)哨兵节点。哨兵模式具有主从复制模式的一切优点,是自动版的主从复制模式。同时,这种模式也有不足之处:①部署困难,难以扩容。②由普通的 Redis 服务器充当哨兵节点,而哨兵节点不参与数据存储,浪费资源。

为了解决哨兵模式需要额外部署哨兵节点、内存扩展困难、数据冗余等问题,从 3.0 版本开始,Redis 推出了 Redis Cluster(本书称 Redis 分片集群)。Redis 分片集群是 Redis 的分布式解决方案。它非常优雅地解决了 Redis 集群方面的问题。Redis 分片集群去掉额外的哨兵节点,利用哈希槽实现数据分片存储,这样极大降低了数据的冗余度,并提高了读写能力。此外,分片集群支持在线扩容。对于分片集群,还存在以下一些问题:①如果一个分片的主从节点都宕机,则整个集群宕机。②分片集群不支持多数据库空间,单机版 Redis 支持使用 16 个数据库,而分片集群模式下每个 Redis 节点只能使用 1 个数据库,即 db 0。③键是数据分区的最小单位,分片集群不能将大键值对象映射到不同节点。④无法解决热点读写问题。分片集群中数据虽然均匀分布,但不能避免请求偏移到某个槽,造成单个槽压力过大。

第 8 章

Redis仓库

Redis 是高性能的 NoSQL 数据库，经常作为缓存应用于各大互联网架构中。同时，它也可以作为数据持久化存储的工具。本章介绍将 Redis 作为数据持久化存储工具（即 Redis 仓库）的使用方法。注意，Redis 仓库要求 Redis 的最低版本为 2.8.0，并且不要使用事务。

8.1 入门程序

利用 Redis 仓库存储对象与 Spring 的 JPA（Java Persistence API，Java 持久层 API）类似。例如，要在 Redis 中存储一个公民及其所在城市的信息，可以按下述步骤实现。

1. 编写实体类

可以创建一个名为 Person 的类，用于封装公民的信息；同时，创建一个名为 Address 的类，用于封装公民的地址信息。Person 类和 Address 类的实例就是域对象，Redis 持久化域对象时使用的数据类型为哈希。其中，Person 类的代码如文件 8-1 所示。

【文件 8-1】 Person.java

```
1   @Data
2   @RedisHash("people")
3   public class Person {
4       @Id
5       String id;
6       String firstname;
7       String lastname;
8       Address address;
9      //此处省略了 toString()方法
10  }
```

如文件 8-1 所示，用于创建域对象（以下称对象）的 Person 类由@Data（lombok.Data）注解标记（第 1 行），表明在安装了 lombok 插件后，可以省去 getters/setters 方法。同时，第 2 行的@RedisHash 和第 4 行的@Id 负责创建持久化哈希的键 key。并且，@Id 注解所标注的属性 id 称为标识符属性。

2. 配置 Redis 仓库

接下来，需要配置 Redis 的连接工厂和操作 Redis 的 RedisTemplate 类。代码如文件 8-2 所示。

【文件 8-2】 DemoConfig.java

```
1    @Configuration
2    @EnableRedisRepositories
3    public class DemoConfig {
4        @Bean
5        public RedisConnectionFactory connectionFactory() {
6            ...
7            return factory;
8        }
9
10       @Bean
11       public RedisTemplate<Person, String>
12         redisTemplate(RedisConnectionFactory redisConnectionFactory) {
13           RedisTemplate<Person, String> template =
14                 new RedisTemplate<Person, String>();
15           template.setConnectionFactory(redisConnectionFactory);
16           return template;
17       }
18   }
```

如文件 8-2 所示，第 1 行用@Configuration 标注 DeomConfig 类，表示该类是一个配置类，被注解的类内部包含一个或多个被@Bean 注解标注的方法，这些方法将会被 AnnotationConfigApplicationContext 类扫描，用于初始化 Spring 容器并创建 Bean。其中，@Bean 是一个方法级别的注解，表示方法产生一个由 Spring 管理的 Bean，默认用方法名作为 Bean 的 id。第 2 行的@EnableRedisRepositories 表示开启 Redis 仓库，该注解允许 Spring 扫描包中的仓库类或接口（见第 3 步），然后将对象持久化到 Redis，而不是传统的关系数据库。第 4 行的@Bean 注解标注的方法用于返回一个 RedisConnectionFactory 对象。本案例采用的 Redis 客户端是 Jedis，创建 RedisConnectionFactory 对象的相关代码可参考 3.2.2 节和 3.2.3 节。

3. 创建仓库接口

接下来创建仓库接口，该接口可配合文件 8-2 中的@EnableRedisRepositories 注解，实现在 Redis 仓库中对 Person 对象进行最基础的增、删、改、查操作，代码如文件 8-3 所示。

【文件 8-3】 PersonRepository.java

```
1    @Repository
2    public interface PersonRepository extends CrudRepository<Person,
3      String> {
4    }
```

其中，CrudRepository 接口定义了以下方法。

(1) T save(T entity);：保存单个实体。

(2) Iterable save(Iterable<? extends T> entities);：保存实体集合。

(3) T findOne(id);：根据标识符属性 id 查找实体。

(4) boolean exists(id);：根据标识符属性 id 判断实体是否存在。

(5) Iterable findAll();：查询所有实体，不用或慎用。

(6) long count();：查询实体数量。

(7) void delete(id);:根据标识符属性 id 删除实体。

(8) void delete(T entity);:删除一个实体。

(9) void delete(Iterable <? extends T > entities);:删除一个实体集合。

4. 编写测试代码

编写测试代码,用于测试 PersonRepository 接口中的 save()、findById()、count()和 delete()方法。测试代码如文件 8-4 所示。

【文件 8-4】 DemoTest.java

```
1   @RunWith(SpringJUnit4ClassRunner.class)
2   @ContextConfiguration(classes = DemoConfig.class)
3   public class DemoTest {
4       @Autowired
5       private PersonRepository repo;
6
7       @Test
8       public void basicCrudOperations() {
9           Person person = new Person();
10          person.setId(UUID.randomUUID().toString());
11          person.setFirstname("John");
12          person.setLastname("Smith");
13          person.setAddress(new Address("emond", "andor"));
14          repo.save(person);
15
16          System.out.println(repo.findById(person.getId()));
17          System.out.println(repo.count());
18          repo.delete(person);
19      }
20  }
```

测试代码运行结果如图 8-1 所示。

```
Optional[Person{id='edfb2d03-ab41-46d4-a490-fafd417755a5',
  firstname='John', lastname='Smith',
 address=Address{id='null', state='emond', city='andor'}}]
1
```

图 8-1 文件 8-4 测试代码运行结果

8.2 对象映射基础

本节介绍 Spring Data 对象映射、对象创建、属性访问的基础知识。注意,本节内容仅适用于不使用底层数据存储(如 JPA)的 Spring Data 模块。

Spring Data 对象映射的核心问题是创建域对象,并将特定于本地存储器的数据结构映射到这些对象上。完成这个任务需要两个基本步骤。

(1) 使用构造方法创建域对象。

(2) 为所有非私有属性赋值。

8.2.1 对象创建

Spring Data 会自动检测实体类的构造方法，用于创建该类型的对象，具体原则如下。

（1）如果有一个用@PersistenceCreator 注解标注的静态工厂方法，则使用该方法。

（2）如果只有一个构造方法，则使用该构造方法。

（3）如果有多个构造方法，并且恰好有一个用@PersistenceConstructor 标注，则使用被该注解标注的构造方法。

（4）如果有无参构造方法，则使用它。其他构造方法将被忽略。

为避免映射带来的开销，Spring Data 在创建对象时做了一个优化：在运行时创建一个默认的工厂类，并由工厂类直接调用对应的实体类的构造方法。例如，实体类 Person 的定义如下：

```
1    class Person {
2      Person(String firstname, String lastname) { … }
3    }
```

Spring Data 在运行时会创建下面的工厂类：

```
1    class PersonObjectInstantiator implements ObjectInstantiator {
2      Object newInstance(Object... args) {
3        return new Person((String) args[0], (String) args[1]);
4      }
5    }
```

为了使实体类能够进行此类优化，它需要遵守以下一组约束。

（1）该类不能定义为私有。

（2）该类不能是非静态的内部类。

（3）该类不能是 CGLib 代理类。

（4）该类提供给 Spring Data 使用的构造方法不能是私有的。

8.2.2 属性赋值

创建域对象后，Spring Data 还要给该对象的所有持久属性赋值。首先对标识符属性（id 属性）赋值，除非该属性已经由实体类的构造方法赋值。这样做是为了解析嵌套对象的引用。之后，对域对象上尚未由构造方法赋值的所有非瞬时（持久）属性进行赋值。为此，Spring Data 使用以下原则。

（1）如果属性是不可变的，但提供了一个 with…方法，则 Spring Data 使用 with…方法创建一个具有新属性值的新域对象。

（2）如果定义了属性访问方法（getters 和 setters 方法），则 Spring Data 将调用 setters 方法完成属性赋值。

（3）如果属性是可变的，则 Spring Data 直接设置属性值。

（4）如果属性是不可变的，则 Spring Data 利用持久化操作使用的构造方法（参见 8.2.1 节）来创建对象的副本。

（5）默认情况下，Spring Data 直接设置属性值。

与对象创建过程类似，Spring Data 使用运行时生成的属性访问器类与域对象交互。例如，对于文件 8-5 的 Person 类，Spring Data 会创建如文件 8-6 所示的属性访问器类。

【文件 8-5】　Person.java

```
1    public class Person {
2      private final Long id;
3      private String firstname;
4      private @AccessType(Type.PROPERTY) String lastname;
5      Person() {
6        this.id = null;
7      }
8      Person(Long id, String firstname, String lastname) {
9        //属性赋值
10     }
11     Person withId(Long id) {
12       return new Person(id, this.firstname, this.lastame);
13     }
14     void setLastname(String lastname) {
15       this.lastname = lastname;
16     }
17   }
```

【文件 8-6】　PersonPropertyAccessor.java

```
1    class PersonPropertyAccessor implements PersistentPropertyAccessor {
2      private static final MethodHandle firstname;
3      private Person person;
4      public void setProperty(PersistentProperty property, Object value) {
5        String name = property.getName();
6        if("firstname".equals(name)) {
7          firstname.invoke(person, (String) value);
8        } else if("id".equals(name)) {
9          this.person = person.withId((Long) value);
10       } else if("lastname".equals(name)) {
11         this.person.setLastname((String) value);
12       }
13     }
14   }
```

如文件 8-6 所示，PersistentPropertyAccessor 接口用于保存对象的持久化属性。默认情况下，Spring Data 使用属性访问方法（getters 或 setters 方法）来读取和写入属性值。根据私有属性的可见性规则，使用 MethodHandle 类与私有属性交互（第 2 行，第 6、7 行）。文件 8-5 中的 Person 类提供了一个 withId()方法（第 11～13 行），用于设置标识符属性的值。例如，当一个域对象保存到 Redis 仓库中并生成了标识符时，调用 withId()方法创建一个新的 Person 对象。后续的所有改变都将在这个新的对象中生效，而之前的域对象则保持不变。第 11 行使用属性访问，可以在不使用 MethodHandle 的情况下直接调用 setters 方法。为了使实体类能够进行此类优化，它需要遵守一组约束。

（1）类不能存放在默认包中或 java 包下。

（2）类及其构造方法必须是公有的（public）。

(3) 内部类的类型必须是静态的(static)。

(4) Java 运行时环境必须允许在原始类加载器中声明类。Java 9 和更新版本会有一些限制。

默认情况下,Spring Data 会首先使用生成的属性访问器,代码如文件 8-7 所示。

【文件 8-7】 Person. java

```
1    package com.example.repository.fundamental.entity;
2    class Person {
3      private final @Id Long id;
4      private final String firstname, lastname;
5      private final LocalDate birthday;
6      private final int age;
7      private String comment;
8      private @AccessType(Type.PROPERTY) String remarks;
9      static Person of(String firstname, String lastname,
10        LocalDate birthday) {
11       return new Person(null, firstname, lastname, birthday,
12         Period.between(birthday, LocalDate.now()).getYears());
13     }
14     Person(Long id, String firstname, String lastname, LocalDate birthday,
15        int age) {
16       this.id = id;
17       this.firstname = firstname;
18       this.lastname = lastname;
19       this.birthday = birthday;
20       this.age = age;
21     }
22     Person withId(Long id) {
23       return new Person(id, this.firstname, this.lastname, this.birthday,
24         this.age);
25     }
26     void setRemarks(String remarks) {
27       this.remarks = remarks;
28     }
29   }
```

如文件 8-7 所示,属性 id 被声明为 final,但在构造方法中设置为 null(第 11 行)。该类定义了一个公有的用于设置标识符的 withId()方法(第 22~25 行)。例如,当一个 Person 域对象保存到 Redis 仓库中并生成了标识符时可以调用该方法。在创建新的 Person 域对象时,原有 Person 域对象保持不变。with…方法是可选的,因为构造方法(第 14~21 行)设置的属性会使用新的标识符创建新的对象。firstname 和 lastname 属性是普通的不可变属性(第 4 行),可利用 getters 方法获取。age 属性是不可变属性,可以从生日属性派生而来。在文件 8-7 中,Spring Data 使用唯一声明的构造方法创建对象。并且,该构造方法也必须将 age 作为参数,否则在设置属性 age 的值时,会由于其不可变且不存在 with…方法而失败。comment 属性是可变的(第 7 行),可以直接赋值。remark 属性是可变的(第 8 行),可以直接设置 remark 属性的值或调用 setters 方法为其赋值。

8.3 对象-哈希映射

Redis仓库支持将对象持久化为哈希。一般由RedisConverter接口(org.springframework.data.redis.core.convert.RedisConverter)完成对象-哈希(Object-to-Hash)转换。默认地,Spring使用RedisConverter将对象属性值与Redis本机byte[]数组建立映射。如8.1节的Person类(见文件8-1),映射后的结果为:

```
1    _class = org.example.Person
2    id = e2c7dcee - b8cd - 4424 - 883e - 736ce564363e
3    firstname = John
4    lastname = Smith
5    address.city = emond
6    address.country = andor
```

其中,_class属性包含在根级别以及任何嵌套接口或抽象类型上(第1行);简单属性值按路径映射(第3、4行);复杂类型的属性通过其点路径进行映射(第5、6行)。具体的映射规则如表8-1所示。

表8-1 对象-哈希映射规则

类 型	示 例	映射后的值
简单类型(如字符串)	String firstname="rand";	firstname="rand"
Byte array(byte[])	byte[]image="rand".getBytes();	image="rand"
复杂类型(如Address)	Address address=new Address("emond");	address.city="emond"
简单类型列表	List<String> names=asList("dragon","therin");	names.[0]="dragon", names.[1]="therin"
简单类型的Map	Map<String,String> atts=asMap({"eye-color","grey"},{"hair-clolor,...}	atts.[eye-color]="grey", atts.[hair-color]="..."
复杂类型列表	List<Address> addresses=asList(new Address("emond"),new Address("..."	addresses.[0].city="emond", addresses.[1].city="..."
复杂类型的Map	Map<String,Address> addresses=asMap({"home",new Address("emond"),{"work",new Address("...")}}	addresses.[home].city="emond", addresses.[work].city="..."

映射行为可以通过在RedisCustomConversions类(org.springframework.data.redis.core.convert.RedisCustomConversions)中注册相应的转换器来定制。这些转换器可以处理字节数组与字节数组及字节数组与Map<String,byte[]>间的相互转换。第一种转换适用于将复杂类型转换为二进制的JSON表示,但仍然使用默认的哈希结构;而第二种转换可以完全控制产生的哈希。将对象写入Redis哈希会删除原有哈希中的内容,并重新创建整个哈希,因此未映射的数据会丢失。以下是两个自定义转换器的例子,代码如文件8-8和文件8-9所示。

【文件8-8】 AddressToBytesConverter.java

```
1    @WritingConverter
2    public class AddressToBytesConverter
```

```
3        implements Converter<Address, byte[]> {
4      private final Jackson2JsonRedisSerializer<Address> serializer;
5      public AddressToBytesConverter() {
6         serializer = new Jackson2JsonRedisSerializer<Address>(
7               Address.class);
8         serializer.setObjectMapper(new ObjectMapper());
9      }
10     @Override
11     public byte[] convert(Address value) {
12        return serializer.serialize(value);
13     }
14  }
```

【文件 8-9】 BytesToAddressConverter.java

```
1   @ReadingConverter
2   public class BytesToAddressConverter
3        implements Converter<byte[], Address> {
4      private final Jackson2JsonRedisSerializer<Address> serializer;
5      public BytesToAddressConverter() {
6         serializer = new Jackson2JsonRedisSerializer<Address>(
7               Address.class);
8         serializer.setObjectMapper(new ObjectMapper());
9      }
10     @Override
11     public Address convert(byte[] value) {
12        return serializer.deserialize(value);
13     }
14  }
```

文件 8-8 和文件 8-9 分别定义了一个写转换器和一个读转换器。这两个转换器分别用@WritingConverter 注解(文件 8-8 的第 1 行)和@ReadingConverter 注解标注(文件 8-9 的第 1 行)。其中的转换器 Jackson2JsonRedisSerializer 可以利用 ObjectMapper 类读取 JSON 数据,并且该转换器可用于绑定类型化 Bean 或非类型化 HashMap 实例。空对象被序列化为空数组,反之亦然。文件 8-8 的 convert()方法(第 11～13 行)执行序列化操作,文件 8-9 的 convert()方法(第 11～13 行)执行反序列化操作。

为了查看序列化和反序列化的结果,可以定义一个名为 Address 的实体类。代码如文件 8-10 所示。

【文件 8-10】 Address.java

```
1   @Data
2   public class Address {
3      @Id
4      private String id;
5      private String state;
6      private String city;
7       //此处省去了 toString()方法
8   }
```

编写测试代码后，只调用 AddressToBytesConverter 的 convert()方法执行序列化，控制台输出的序列化结果(二进制数据的十进制表示)如图 8-2 所示。

对上述数据再次执行反序列化，即调用 BytesToAddressConverter 的 convert()方法，控制台输出如图 8-3 所示。

```
C:\software\jdk-11.0.13\bin\java.exe ...
[123, 34, 105, 100, 34, 58, 110, 117, 108, 108, 44, 34, 115, 116, 97,
 116, 101, 34, 58, 34, 97, 110, 100, 111, 114, 34, 44, 34, 99, 105,
 116, 121, 34, 58, 34, 101, 109, 111, 110, 100, 34, 125]

Process finished with exit code 0
```

图 8-2 序列化后的结果

```
C:\software\jdk-11.0.13\bin\java.exe ...
Address{id='null', state='andor', city='emond'}
```

图 8-3 反序列化后的结果

8.4 键空间

键空间用于定义 Redis 哈希所创建的实际键的前缀。默认情况下，前缀设置为 Object #getClass().getName()。这个默认值可以通过@RedisHash 注解设置或通过编程来更改。但是，使用注解设置的键空间会取代任何其他设置。

例如，文件 8-1 中，@RedisHash 注解已将 Person 类的 Redis 哈希键的前缀设置为 people(第 2 行)。如果要修改此设置，可以通过设置键空间的方式实现。具体步骤如下。

1. 创建配置类

创建一个名为 MyKeyspaceConfig 的配置类，利用 KeyspaceConfiguration 类(org.springframework.data.redis.core.convert.KeyspaceConfiguration)设置键空间和生存时间选项，代码如文件 8-11 所示。

【文件 8-11】 MyKeyspaceConfig.java

```
1    @Configuration
2    @EnableRedisRepositories
3    public class MyKeyspaceConfig {
4      ...//RedisConnectionFactory 和 RedisTemplate Bean 定义,略
5      @Bean
6      public RedisMappingContext keyValueMappingContext() {
7        return new RedisMappingContext(new MappingConfiguration(
8              new IndexConfiguration(), new MyKeyspaceConfiguration()));
9      }
10
11     public static class MyKeyspaceConfiguration
12        extends KeyspaceConfiguration {
13       protected Iterable<KeyspaceSettings> initialConfiguration() {
14         return Collections.singleton(new KeyspaceSettings(
15            Person.class, "people"));
16       }
17     }
18   }
```

文件 8-11 采用的是编程方式将 Person 类的 Redis 哈希键的前缀设置为 people，也可以利用@EnableRedisRepositories 实现同样的效果，代码如文件 8-12 所示。

【文件 8-12】 KeyspaceConfig. java

```
1   @Configuration
2   @EnableRedisRepositories(keyspaceConfiguration =
3           KeyspaceConfig.MyKeyspaceConfiguration.class)
4   public class KeyspaceConfig {
5       ...//RedisConnectionFactory 和 RedisTemplate Bean 定义,略
6       public static class MyKeyspaceConfiguration
7               extends KeyspaceConfiguration {
8           protected Iterable<KeyspaceSettings> initialConfiguration() {
9               return Collections.singleton(
10                      new KeyspaceSettings(Person.class, "people"));
11          }
12      }
13  }
```

2. 运行测试

当 Person 类由@RedisHash 注解指定了键的前缀后，实际存储的键的前缀以该注解指定的值为准。当@RedisHash 注解没有指定键的前缀时，文件 8-11 或文件 8-12 中指定的前缀才会起作用。为验证这一结论，可以将@RedisHash 注解指定的 Person 类的键前缀删除，并通过文件 8-3 的 PersonRepository 接口将一个 Person 对象存入 Redis。随后，借助 RedisInsight 工具或在 Redis 客户端执行 KEYS * 命令检查键的前缀。程序运行后，通过 RedisInsight 检查的结果如图 8-4 所示。

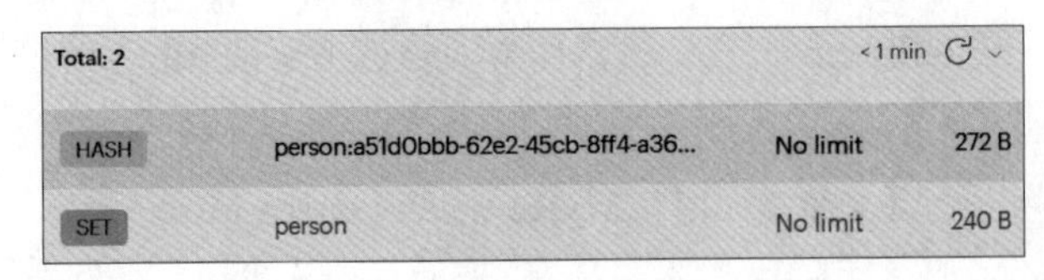

图 8-4 通过 RedisInsight 检查的结果

8.5 辅助索引

作为一个数据持久化工具和缓存，Redis 也是依靠索引提高数据检索速度的。Redis 提供了一个外部键值壳(Shell)。在 API 级别，Redis 只提供主键访问，即利用键查找数据。但是，Redis 并不完全是一个键值存储器，因为值可以是复杂的数据结构。为便于查找复杂结构的数据，Redis 又可以创建不同类型的辅助索引(Secondary Indexes)，包括复合(多列)索引。每次保存数据时，辅助索引值都会写入相应的索引，在对象被删除或过期时删除辅助索引。

本节介绍简单属性索引、地理空间索引及自定义辅助索引。

8.5.1 简单属性索引

对于普通类来说，可以通过@Indexed 注解利用相关属性创建一个辅助索引。例如，对于文件 8-1 的 Person 类，可以利用属性 firstname 创建索引，代码如文件 8-13 所示。

【文件 8-13】 Person.java

```
1   @RedisHash("people")
2   public class Person {
3       private @Id String id;
4       private @Indexed String firstname;
5       private String lastname;
6       private Address address;
7         //此处省略了构造方法和 toString()方法
8   }
```

随后,为了在 Redis 中保存 Person 对象,可以创建一个 PersonRepository 接口,代码同文件 8-3。索引针对真实的属性值建立。向 Redis 中保存两个 Person 对象,如 Aviendha 和 Thatcher,会建立索引,如图 8-5 第 4)行和第 7)行所示。

```
127.0.0.1:6379> keys *
1) "people"
2) "people:firstname:Thatcher"
3) "people:firstname:Aviendha"
4) "people:7a3c14ae-cf98-4efe-85d0-09e19690cda1:idx"
5) "people:48c7d9e9-6df6-4e70-8c21-58cf177ee388"
6) "people:7a3c14ae-cf98-4efe-85d0-09e19690cda1"
7) "people:48c7d9e9-6df6-4e70-8c21-58cf177ee388:idx"
127.0.0.1:6379>
```

图 8-5 建立的索引

此外,可以利用键空间通过编程方式配置索引,此时无须在属性上用@Indexed 注解标注。配置类代码如文件 8-14 所示。

【文件 8-14】 IndexingConfig.java

```
1   @Configuration
2   @EnableRedisRepositories(indexConfiguration = MyIndexConfig.class)
3   public class IndexingConfig {
4     //此处省略了 Redis 连接工厂和 RedisTemplate Bean 的定义
5     public static class MyIndexConfig extends IndexConfiguration {
6       @Override
7       protected Iterable<IndexDefinition> initialConfiguration() {
8         return Collections.singleton(new SimpleIndexDefinition("people",
9             "firstname"));
10      }
11    }
12  }
```

8.5.2 地理空间索引

Redis 地理空间索引利用一个有序集合,使用地理哈希技术按纬度和经度对位置进行索引。有序集合的分数表示经度和纬度的交替位,因此可以将有序集合的线性分数映射为地球表面的许多小方块。借助 8-邻域搜索算法,可以实现按半径搜索元素。

假设文件 8-10 中的 Address 类中包含 Point 类型的位置属性,该属性保存特定地址的地理坐标。通过用@GeoIndexed 注解标注相关属性,Spring Data Redis 可以使用 Redis GEO 命令添加相关值。修改后的 Address 类如文件 8-15 所示。

【文件 8-15】 Address.java

```
1    public class Address {
2        @Id
3        private String id;
4        private String state;
5        private String city;
6        private @GeoIndexed Point location;
7        //此处省略了构造方法和 toString()方法
8    }
```

随后，可以编写测试代码，分别对 Person 类和 Address 类的相关属性赋值，并将 Person 对象保存到 Redis 哈希中。部分测试代码如文件 8-16 所示。

【文件 8-16】 部分测试代码

```
1    @Test
2    public void testGeoIndexing() {
3        Person rand = new Person("rand", "al'thor");
4        rand.setAddress(new Address("DanFou","NewZed", new Point(
5          13.361389D, 38.115556D)));
6        repository.save(rand);
7    }
```

运行测试代码后，可通过 Redis 客户端查看索引的建立情况，结果如图 8-6 所示。

```
127.0.0.1:6379> keys *
1) "people"
2) "people:3f872562-f43b-4f1e-8d7e-5b3f389ddd83:idx"
3) "people:3f872562-f43b-4f1e-8d7e-5b3f389ddd83"
4) "people:address:location"
```

图 8-6 建立的地理空间索引

8.5.3 自定义辅助索引

Redis 中，最简单的自定义辅助索引就是有序集合的分数。在有序集合中，元素都是按照分数由低到高排列的。因此，可以利用分数作为索引，快速检索数据。例如，可以根据用户给定的价格区间快速检索对应的商品。

【例 8-1】 现有 4 件商品 A、B、C、D。它们的价格分别为 169.0 元、139.0 元、399.0 元、218.0 元。根据用户要求，计算价格为 100～200 元的商品的数量，并输出商品的名称和价格。测试代码如文件 8-17 所示。

【文件 8-17】 例 8-1 测试代码

```
1    @Test
2    public void testZSortIndexing(){
3        ZSetOperations<String, String> ops = template.opsForZSet();
4        ops.add("myindex","A",169.0);
5        ops.add("myindex","B", 139.0);
6        ops.add("myindex","C", 399.0);
7        ops.add("myindex","D", 218.0);
8        System.out.println(ops.count("myindex",100,200));
9        Set<ZSetOperations.TypedTuple<String>> p =
```

```
10            ops.rangeByScoreWithScores("myindex",100,200);
11        p.forEach(item -> System.out.println("name:" + item.getValue()
12            + " price: " + item.getScore()));
13    }
```

如文件 8-17 所示，第 4～7 行将数据写入有序集合。数据写入有序集合后，将按照分数（本例为价格）由低到高排序，以分数建立辅助索引。第 8 行计算有序集合 myindex 中价格在 100～200 元的商品的数量，其中 count()方法可以在不检索该范围（100～200 元）元素的情况下计算该范围内元素的数量。第 9、10 行调用 ZSetOperations 接口的 rangeByScoreWithScores()方法，从给定的有序集合 myindex 中按照辅助索引（本例为价格）检索指定范围内的商品，第 11、12 行输出检索到的商品名称和价格。

在社交网络（或电商系统）中，常常有这样的需求：按照作品（或商品）的点赞数（或浏览量）生成一个排行榜。为了从排行榜中快速检索对应的作品（或商品），可以自定义一个辅助索引。借助辅助索引，可以从排行榜中快速链接到作品（或商品）的详情信息。这样，就需要两种数据类型共同参与：对于作品（或商品）的详情信息，可以定义为哈希类型；对于排行榜，可以采用有序集合存储。而这两种类型的数据之间的关系可由图 8-7 来描述。

哈希类型

键	值
	id:1;name:n_1;vote:v_1,...
	…
	id:n;name:n_n;vote:v_n,...

有序集合

键	值	分数
	id:2	v_2
	…	…
	id:i	v_i

图 8-7 两种类型的数据之间的关系

如图 8-7 所示，作品（或商品）的详情信息定义为哈希类型，哈希类型中的字段 vote 代表作品（或商品）的点赞数（或浏览量）。而有序集合中存放的是作品（或商品）的标识符属性（id 属性）的值，元素的分数就是字段 vote 的值。这样，在有序集合中就可以对作品（或商品）按点赞数（或浏览量）排序，生成排行榜。而要通过排行榜查看对应作品（或商品）的详情，则可通过有序集合中的元素 id 的值到哈希中查找对应的数据。对于存放对象的哈希来讲，有序集合中的元素 id 就可以认为是自定义的辅助索引。

【例 8-2】 社交网络中现有 4 位作者提供了 4 个作品，如表 8-2 所示。现要求输出点赞数高于 60 的作品的详情。实现步骤如下。

表 8-2 作者与作品对照表

作品标识符	名 称	标 题	作 者
1	Java	Title_A	ZhaoYun
2	Python	Title_B	HuangZhong
3	Redis	Title_C	MaChao
4	MySQL	Title_D	ZhangFei

1. 创建实体类

创建一个用于封装作品信息的实体类，代码如文件 8-18 所示。

【文件 8-18】 Works. java

```
1   @Data
2   public class Works {
3       private Integer id;            //作品标识符
4       private String name;           //作品名称
5       private String title;          //作品标题
6       private String author;         //作者
7       private Long vote;             //作品获赞数
8   }
```

2. 配置 RedisTemplate 类

修改配置类 IndexingConfig（见文件 8-14），添加对 RedisTemplate 类的配置。可参考文件 3-6 及文件 3-61 或查看本书附带的源代码，此处不再赘述。

3. 编写测试代码

测试代码如文件 8-19 所示。

【文件 8-19】 例 8-2 测试代码

```
1   @Test
2   public void testIndexing2(){
3       String[] name = {"Java","Python","Redis","MySQL"};
4       String[] title = {"Title_A","Title_B","Title_C","Title_D"};
5       String[] author = {"ZhaoYun","HuangZhong","MaChao","ZhangFei"};
6
7       HashOperations<String, String, Object> hashOperations =
8           template.opsForHash();
9       ZSetOperations<String, Object> zSetOperations =
10          template.opsForZSet();
11      for(int i = 0;i < name.length;i++){
12          Works work = new Works();
13          work.setId(i);
14          work.setAuthor(author[i]);
15          work.setName(name[i]);
16          work.setTitle(title[i]);
17          work.setVote(Math.round((Math.random() * 100 + 1)));
18          hashOperations.put("work",String.valueOf(i),work);
19          zSetOperations.add("works_vote",String.valueOf(work.getId()),
20              work.getVote());
21      }
22      Set<ZSetOperations.TypedTuple<Object>> set = zSetOperations
23          .rangeByScoreWithScores("works_vote",60,100);
24      set.forEach(s -> System.out.println(
25          hashOperations.get("work",s.getValue())));
26  }
```

如文件 8-19 所示，第 12～17 行进行数据初始化。其中第 17 行用 1～99 的随机数模拟作品获赞数。第 18 行将作品基础数据信息以 work 为键保存到 Redis 哈希中。第 19 行将作品 id 值（标识符属性的值）保存到有序集合 works_vote 中，元素的分数即为作品获赞数。第 22、23 行从有序集合中取出满足要求的元素的 id。第 24、25 行根据检索到的元素的 id

值在以 work 为键的哈希中查找对应的值并输出。测试代码运行结果如图 8-8 所示。同时，可以用 RedisInsight 查看 Redis 中保存的数据，如图 8-9 所示。

```
Works(id=1, name=Python, title=Title_B, author=HuangZhong, vote=60)
```

图 8-8　文件 8-19 测试代码运行结果

SORTED SET　works_vote

Key Size: 83 B　Length: 4　TTL: No limit

Member	Score ↑
0	31
3	31
2	49
1	60

HASH　work

Key Size: 911 B　Length: 4　TTL: No limit　Last refresh: < 1 min

Field	Value
2	{"@class":"com.example.repository.indexing.simple.entity.Works","i
0	{"@class":"com.example.repository.indexing.simple.entity.Works","i
3	{"@class":"com.example.repository.indexing.simple.entity.Works","i e_D","author":"ZhangFei","vote":["java.lang.Long",31]}
1	{"@class":"com.example.repository.indexing.simple.entity.Works", "id":1,"name":"Python","title":"Title_B","author":"HuangZhong", "vote":["java.lang.Long",60]}

图 8-9　Redis 中保存的数据

8.6　查询

8.6.1　示例查询

示例查询(Query by Example，QBE)是一种用户友好的查询技术，查询接口简单。它是一种通过实体对象设置查询条件的值，进而执行查询的方法。它允许动态创建查询，不需要编写包含字段名的查询。示例查询不要求使用特定于存储介质的查询语言编写查询程序。

示例查询 API 由以下三部分组成：

(1) Probe(探针)：具有已赋值属性的对象实例。

(2) ExampleMatcher(示例匹配器)：包含有关如何匹配特定字段(属性)的详细信息，它可以在多个示例中重复使用。

(3) Example(示例)：由探针和示例匹配器组成，于创建查询。

示例查询适合的场景：①使用一组静态或动态约束查询数据集。②需要频繁重构域对象，而不用担心破坏现有的查询。③与底层数据存储 API 无关的工作。同时，示例查询也有限制：①不支持嵌套或分组的属性约束，例如 firstname＝?0 or(firstname＝?1 and lastname＝?2)。②仅支持字符串的 starts/contains/ends/regex 匹配和其他属性类型的精确匹配。

例如，对于文件 8-1 的 Person 类，可以采用两种方式执行示例查询：第一，使用 Example 接口(org. springframework. data. domain. Example)的工厂方法 of()创建 Example 实例；第二，使用 ExampleMatcher 接口(org. springframework. data. domain. ExampleMatcher)创建 Example 实例。

方法一：

```
1    Person person = new Person();
2    person.setFirstname("Dave");
3    Example<Person> example = Example.of(person);
```

方法二：

```
1    Person person = new Person();
2    person.setFirstname("Dave");
3    ExampleMatcher matcher = ExampleMatcher.matching()
4      .withIgnorePaths("lastname")
5      .withIncludeNullValues()
6      .withStringMatcher(StringMatcher.ENDING);
7    Example<Person> example = Example.of(person, matcher);
```

其中，方法二的第3行用于创建一个ExampleMatcher对象，它可以匹配所有的值。第4～6行用于设置查询条件，如忽略lastname属性（第4行）、可包含空值（第5行），并执行后缀字符串匹配（第6行）。默认地，ExampleMatcher的matching()方法执行的是精确匹配查询。如果要执行模糊查询可以使用ExampleMatcher接口的matchingAny()方法。设置查询条件时，可以指定单个属性的值（如，firstname和lastname或者嵌套属性address.city），可以用它们调节匹配条件或者大小写，如下代码所示：

```
1    ExampleMatcher matcher = ExampleMatcher.matching()
2      .withMatcher("firstname", endsWith())
3      .withMatcher("lastname", startsWith().ignoreCase());
4    }
```

可以使用仓库接口运行示例查询。为此，需要自定义一个仓库接口继承org.springframework.data.repository.query.QueryByExampleExecutor接口，该接口规定了用于执行示例查询的一系列方法，如下代码所示：

```
1    public interface QueryByExampleExecutor<T> {
2      <S extends T> Optional findOne(Example<S> example);
3      <S extends T> Iterable<S> findAll(Example<S> example);
4      //此处省略了其他方法
5    }
```

Redis要求利用辅助索引支持Spring Data的示例查询特性。特别是仅使用精确、区分大小写和非空值来构造查询的情况。辅助索引使用集合操作（并、交）来确定匹配的关键字。利用未设置索引的属性执行示例查询不会返回结果，因为不存在索引。示例查询支持检查索引配置，目的仅仅是用来包含查询中索引所覆盖的属性。这样做是为了防止意外包含非索引属性。如对于文件8-1的Person类，根据指定的firstname属性值执行示例查询，可以使用下面的代码：

```
1    Person person = new Person();
2    person.setFirstname("Aviendha");
3    ExampleMatcher matcher = ExampleMatcher.matching()
4      .withStringMatcher(ExampleMatcher.StringMatcher.EXACT);
```

```
5    Example<Person> example = Example.of(person, matcher);
6    Iterable<Person> pers = qy.findAll(example);
```

此时,要查找 firstname 属性为 Aviendha 的记录。利用示例查询在 Redis 中检索数据时,要求利用 firstname 属性创建辅助索引,即将 firstname 属性的定义修改为:

```
@Indexed String firstname;
```

8.6.2 方法查询

方法查询允许利用方法名称自动派生简单的查询。对于文件 8-1 的 Person 类,根据 firstname 属性的值查找相关人员,可以执行如下的方法查询:

```
1    public interface PersonRepo extends CrudRepository<Person, String> {
2      List<Person> findByFirstname(String firstname);
3    }
```

执行方法查询的注意事项如下。

(1) 查询方法中用到的属性已经建立辅助索引。

(2) Redis 仓库的方法查询仅支持对实体和带有分页的实体集合的查询。

Redis 仓库允许采用多种方法来定义查询结果的排序。当检索哈希或集合时,Redis 本身不支持查询中排序(In-Flight Sorting)。因此,方法查询会构造一个比较器(Comparator),在将查询结果作为 java.util.List 返回之前将其应用于查询结果。例如:

```
1    interface PersonRepository extends RedisRepository<Person, String> {
2      List<Person> findByFirstnameOrderByAgeDesc(String firstname);
3      List<Person> findByFirstname(String firstname, Sort sort);
4    }
```

其中,第 2 行定义了一个从方法名称派生的静态排序;第 3 行使用方法参数进行动态排序。

8.7 生存时间

存储在 Redis 中的对象可能仅在一定时间内有效。这在 Redis 中短期持久化对象时尤其有用,因为无须在它们到达生命终点时手动删除它们。可以使用@RedisHash(timeToLive=...)注解或键空间设置以秒为单位的生存时间。

此外,在数值型属性或方法上使用@TimeToLive 注解,可以灵活设置生存时间。但是,不要在同一个类的方法和属性上同时应用@TimeToLive 注解。文件 8-20 展示了属性和方法上的@TimeToLive 注解。

【文件 8-20】 TimeToLiveOnProperty.java

```
1    public class TimeToLiveOnProperty {
2      @Id
3      private String id;
4
```

```
5     @TimeToLive
6     private Long expiration;
7   }
8
9   public class TimeToLiveOnMethod {
10    @Id
11    private String id;
12
13    @TimeToLive
14    public long getTimeToLive() {
15      return new Random().nextLong();
16    }
17  }
```

如果用@TimeToLive显式地标注属性，则该属性会从Redis中读取实际TTL值。-1表示该对象没有关联生存时间。Redis仓库通过RedisMessageListenerContainer订阅Redis键空间的通知。当生存时间设置为一个正数时，相应的EXPIRE命令会被执行。除了保留原始副本外，还会在Redis中保留一个虚拟副本，并设置为在原始副本之后5分钟过期。这样做是为了使Redis仓库能够发布RedisKeyExpiredEvent(键过期事件)。每当键过期时，将过期值保存在Spring的ApplicationEventPublisher中，即使原始值已经被删除，使用Redis仓库的所有应用程序都会收到键过期事件。

默认情况下，应用程序初始化时禁用键过期事件侦听器。可以利用@EnableRedisRepositories注解或RedisKeyValueAdapter类调整启动模式，以启动应用程序的键过期事件侦听器，或者在首次插入具有TTL的实体时启动侦听器。

此外，还有以下几点注意事项。

(1) 延迟或禁用键过期事件侦听器启动会影响键过期事件发布。键过期事件对象会保存过期域对象的副本以及键。禁用的事件侦听器不会发布过期事件。由于侦听器初始化的延迟，延迟期间可能会导致键过期事件的丢失。

(2) 在Redis的配置文件中如果没有设置notify-keyspace-events属性，键空间通知消息侦听器会更改Redis中该属性的设置。现有设置不会被覆盖，因此必须正确设置该属性(或将其留空)。注意，AWS ElastiCache上禁用了CONFIG命令，启用监听器会导致错误。要解决此问题，可将keyspaceNotificationsConfigParameter参数设置为空字符串，这样可以阻止CONFIG命令的使用。

(3) Redis发布-订阅消息不是持久化的。如果键在应用程序关闭时过期，则过期事件不会被处理，这可能导致辅助索引包含对过期对象的引用。

(4) @EnableKeyspaceEvents(shadowCopy=OFF)注解可禁用虚拟副本的存储，并减少Redis中的数据大小。键过期事件对象中将仅包含过期键的ID。

8.8 持久化

8.8.1 持久化引用

使用@Reference注解标记的属性可以存储一个简单的键引用，而不是将值复制到

Redis 中。从 Redis 加载时，引用会自动解析并映射为对象。对于文件 8-1 的 Person 类，可以将@Reference 注解标注到 Address 属性上，并相应修改 Address 类。修改后的 Address 类如文件 8-21 所示。

【文件 8-21】 Address. java

```
1   @RedisHash("add")
2   public class Address {
3       @Id
4       private String id;
5       private String state;
6       private String city;
7       public Address(String id, String state, String city) {
8           this.id = id;
9           this.state = state;
10          this.city = city;
11      }
12      //此处省略了 toString()方法
13  }
```

为将 Person 类(文件 8-1)的对象保存到 Redis 中，还需要编写测试代码，如文件 8-22 所示。

【文件 8-22】 PersistingTest. java

```
1   @RunWith(SpringJUnit4ClassRunner.class)
2   @ContextConfiguration(classes = PersistingConfig.class)
3   public class PersistingTest {
4       @Autowired
5       private PersistingRepo repo;
6
7       @Test
8       public void testPersisting(){
9           Person p = new Person();
10          String id = UUID.randomUUID().toString();
11          p.setId(id);
12          p.setFirstname("Aviendha");
13          p.setLastname("Rand");
14          p.setAddress(new Address(UUID.randomUUID().toString(),"Emond",
15              "Andor"));
16          repo.save(p);
17          System.out.println(repo.findAll());
18      }
19  }
```

随后，可以利用 RedisInsight 工具查看 Redis 中的存储情况，可以看出，Redis 中并没有存放 Address 对象的属性，而是存放了一个 Address 对象的键引用，如图 8-10 所示。

8.8.2 持续部分更新

在某些情况下，如果只需为实体类中的某个属性设置新值，无须加载和重写整个实体类。当准备更改一个属性值时，可借助最后一次活跃时间的会话时间戳来判定。Spring Data Redis 中的 PartialUpdate 类(org. springframework. data. redis. core. PartialUpdate)允许在现有对象上定义赋值和删除操作，同时可以更新实体本身和索引的生存时间。

Field	Value
_class	com.example.repository.persisti...
address	add:567731c1-db2b-4f44-939c-e286b63d3024
firstname	Aviendha
id	6f696f7b-7fb5-4ed9-877e-6f2aa...
lastname	Rand

图 8-10　Redis 中的存储情况

对于文件 8-1 的 Person 类，可以执行部分更新。例如，为 Person 类增加一个属性 age，定义如下：

```
private String age;
```

随后，运行如文件 8-23 所示的测试代码，对 Redis 中的 Person 对象进行部分更新。

【文件 8-23】 PersistingTest. java

```
1     @Test
2     public void testPartialUpdate(){
3        PartialUpdate<Person> update = new PartialUpdate<>(id,Person.class)
4               .set("firstname", "mat")
5               .set("address.city", "emond's field")
6               .del("age")
7               .refreshTtl(true)
8               .set("expiration", 10);
9         template.update(update);
10    }
```

其中，第 4 行和第 5 行重新设置了 firstname 属性和 address 属性的 city 部分；第 6 行删除了 age 属性；第 7 行在 Person 类利用@RedisHash(timeToLive=...)注解设置了生存时间的前提下，开启了更改 TTL 的标记，并在第 8 行设置新的 TTL 值为 10。执行更改前，Redis 中保存的 Person 对象的情况如图 8-11 所示，执行更改后的 Person 对象如图 8-12 所示。由于更改了 TTL，新的 Person 对象将在执行更改 10 秒后消失。

Field	Value
_class	com.example.repository.upd...
address.city	Andor
address.id	d79b0600-b95ɔ-428e-869e-...
address.state	Emond
age	25
firstname	Aviendha
id	54f632ab-8ecc-4349-a743-f0...
lastname	Rand

图 8-11　更改前 Redis 中存储的 Person 对象

Field	Value
_class	com.example.repository.upd...
address.city	emond's field
address.id	8bf317a1-27e1-4415-a7e0-bbc...
address.state	Emond
firstname	mat
id	80ade5de-c12f-46af-9df3-2e...
lastname	Rand
expiration	10

图 8-12　更改后 Redis 中存储的 Person 对象

8.9 Redis 数据仓库集群

可以在 Redis 集群环境中使用 Redis 仓库支持。有关 ConnectionFactory 配置的详细信息，可参阅第 7 章的相关部分。尽管如此，还必须进行一些额外的配置，因为默认的键分布将实体和辅助索引分布到整个集群及其槽中。表 8-3 显示了 Redis 集群上数据的详细信息。

表 8-3 Redis 集群上数据的详细信息

键	类型	槽	节　点
people:e2c7dcee-b8cd-4424-883e-736ce564363e	id for hash	15 171	127.0.0.1:7381
people:a9d4b3a0-50d3-4538-a2fc-f7fc2581ee56	id for hash	7373	127.0.0.1:7380
people:firstname:rand	index	1700	127.0.0.1:7379

某些命令(如 SINTER 和 SUNION)只能在所有相关的键映射到同一槽时在服务器端处理；否则，必须在客户端进行计算。因此，将键空间固定到一个槽非常有用，这样就可以立即利用 Redis 服务器端执行计算。表 8-4 显示了执行此操作时发生的情况(注意槽列中的变化和节点列中的端口值)。

表 8-4 键空间固定后的情况

键	类型	槽	节　点
{people}:e2c7dcee-b8cd-4424-883e-736ce564363e	id for hash	2399	127.0.0.1:7379
{people}:a9d4b3a0-50d3-4538-a2fc-f7fc2581ee56	id for hash	2399	127.0.0.1:7379
{people}:firstname:rand	index	2399	127.0.0.1:7379

当使用 Redis 集群时，可以使用@RedisHash("{yourkeyspace}")注解定义键空间并将其固定到特定的插槽。

8.10 小结

作为一个非关系数据库，Redis 除了作为缓存外，还可以作为一个数据持久化工具来使用。本章介绍了 Redis 作为数据持久化工具——Redis 仓库的基本使用方法。在 Redis 仓库中，采用哈希类型存储持久化数据。因此，在数据持久化时，需要执行对象-哈希映射。对于哈希类型中的键，可以通过程序或注解来设定，还可以利用注解来指定域对象或属性的生存时间。对于哈希类型中的值，持久化时称其为域对象。对于域对象中的引用类型，可以通过设置辅助索引的方式引用域对象中的引用类型(域对象的域对象)。此外，Redis 仓库还提供了自定义查询(示例查询和方法查询)。与作为缓存的工作机制类似，Redis 仓库也支持集群模式。

第 9 章

综合案例

随着互联网的发展，网络购物已成为人们日常消费中不可或缺的部分。近年来，从各大电商平台推出种类繁多的购物促销活动到春节抢红包，再到12306抢火车票，热门商品“秒杀”的场景处处可见。从技术角度来讲，秒杀场景（以下简称秒杀）就是在同一时刻有大量的请求争抢购买同一个商品并完成交易，本质上属于短时突发性高并发访问问题。秒杀的业务特点可概括如下：

（1）定时触发，网络流量在瞬间激增。

（2）大量的秒杀请求中常常只有少部分能够成功。

（3）参与秒杀的商品数量有限，不能超卖，可接受少卖。

（4）不要求立即返回真实的订单结果。

本章以一个简单的秒杀系统为例，介绍Redis在高并发场景下的应用。本案例使用Spring Boot、MyBatis-Plus构建Web应用，采用MySQL数据库存储数据。同时，使用Redis缓存、RabbitMQ消息队列来处理高并发场景。

9.1 系统架构设计

秒杀的核心业务是一个商品售卖系统，图9-1从用户角度展示了秒杀操作的基本流程。

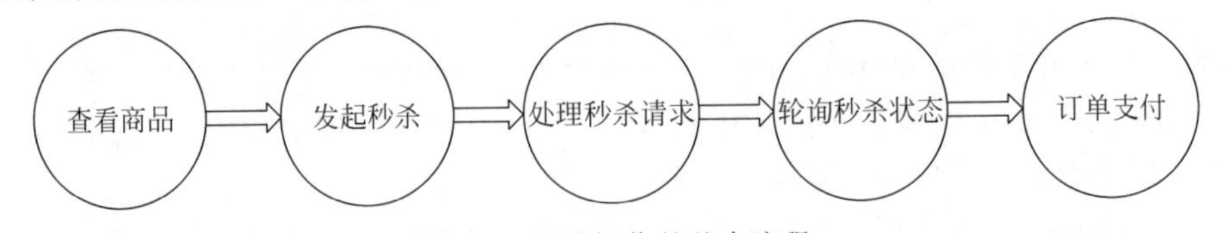

图9-1 秒杀操作的基本流程

虽然秒杀操作的流程简单，但秒杀系统是一个超大流量的并发读写系统，对系统的性能、并发处理能力和可用性都有很高的要求。因此，秒杀系统应该与现有的商品售卖系统分离，作为一个独立的子系统进行设计。

首先，作为一个商品售卖系统，它应具备基础的商品售卖功能。因此，涉及用户、商品管理、库存管理、订单及支付管理等功能。功能架构设计方案如图9-2所示。

其次，秒杀系统总要面对访问量激增的情况，它应具备处理大量并发请求的能力。为及时处理大量的并发请求，提升吞吐量，应在服务器端进行负载均衡。负载均衡架构设计方案如图9-3所示。

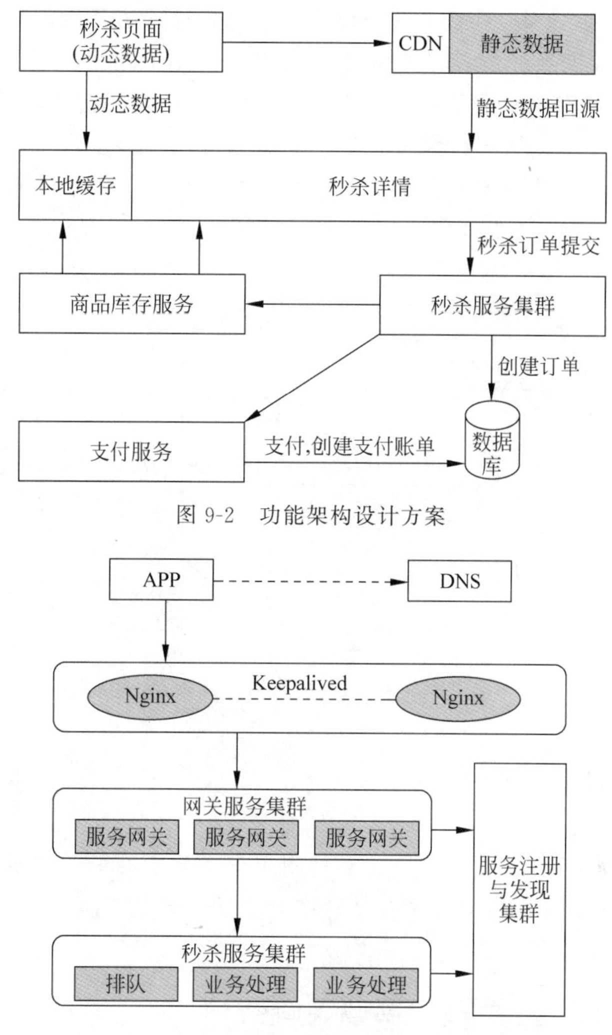

图 9-2　功能架构设计方案

图 9-3　负载均衡架构设计方案

如图 9-3 所示,Nginx 是目前得到广泛应用的 HTTP 服务器和 Web 反向代理服务器,它具有占用内存少、支持高并发请求等特点。通常来说,一个正常 Nginx Linux 服务器的请求处理能力可以达到 500 000～600 000 次/s。同时,Nginx 服务器还具有负载均衡的能力,可以将大量请求均匀分发到不同的服务网关进行处理。为保证秒杀服务的可用性,图 9-3 的设计方案中提供了 Nginx 的集群配置。即由两台 Nginx 提供服务,并且用 Keepalived 软件监控 Nginx 服务器的状态,如果有一台 Nginx 服务器宕机,或出现故障,Keepalived 将检测到,并将有故障的服务器从系统中剔除,同时使用其他服务器代替该服务器的工作。当服务器工作正常后,Keepalived 自动将服务器加入到服务器群中,这些工作全部自动完成,不需要人工干预,需要人工做的只是修复故障的服务器。

对于采用微服务架构的秒杀系统,还需要配置服务网关。图 9-3 给出的设计方案中,为了提升系统可用性,配置了服务网关集群。同时,为了减轻服务器端的压力,对到达的并发请求实行排队限流。

再次,为缩短秒杀系统的响应时间,缓存技术是必不可少的。缓存架构设计方案如

图 9-4 所示。其中，Nginx 缓存用于存储商品图片，以应对秒杀开始前大量的商品详情查询请求；Redis 用于缓存商品详情、商品库存和秒杀状态；APP 端可以缓存商品详情及用户登录信息。MySQL 数据库用于记录秒杀成功后的订单信息，为提高可用性，可采用主从结构的 MySQL 集群。MySQL 从节点作为主节点的数据备份节点。当 MySQL 主节点出现故障时，可以由从节点继续提供服务。同时，主从架构的 MySQL 集群还具备负载均衡的能力，可以由多个节点分摊读写压力。

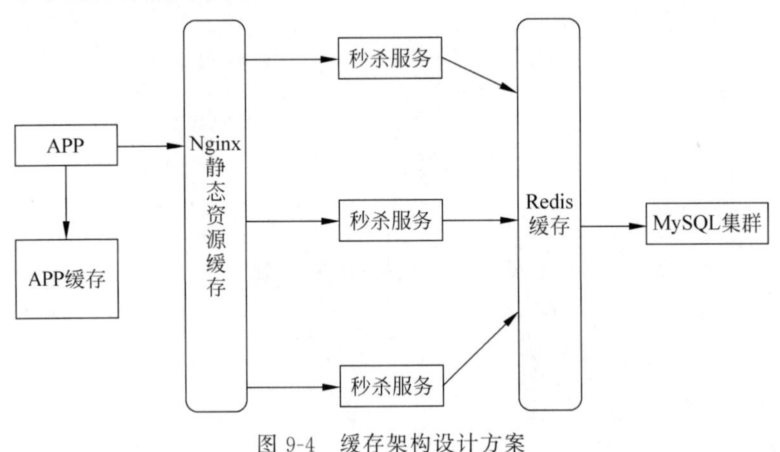

图 9-4　缓存架构设计方案

9.2　简单的售卖系统

本节要设计的简单售卖系统，其功能就是用户登录后查看商品列表（本节以华为 Mate 100 手机和 Mate 100 Pro 手机为待售商品），在秒杀活动开始后，已登录的用户可参与相关商品的秒杀活动，并生成秒杀订单。

9.2.1　简单的售卖系统实现功能

1. 查看参与秒杀活动的商品信息

本系统假设参与售卖的商品为华为 Mate 100 手机和 Mate 100 Pro 手机。为有效跟踪订单，需要参与购买的用户提前注册。当用户登录后，可以查看待售商品列表，该列表列举待售商品的名称、图片、库存数量、售卖价格等信息。随后，可在商品详情页查看详细的商品信息。商品列表页和商品详情页分别如图 9-5 和图 9-6 所示。

秒杀商品列表

商品名称	商品图片	商品原价	秒杀价	库存数量	详情
Mate 100		6299.00	669.00	10	详情
Mate 100 Pro		9299.00	929.00	9	详情

图 9-5　商品列表页

图 9-6 商品详情页

2. 具有秒杀功能

本章以当前设计的简单的商品售卖系统为案例,最终将其扩展为一个商品的秒杀系统。因此,该商品售卖系统应具备商品秒杀功能。具体为,可以事先设置秒杀活动的开始和结束时间。秒杀活动未开始时,"秒杀"按钮显灰色(不可用);只有当秒杀活动开始后,"秒杀"按钮才可以使用。为公平起见,规定每位登录用户只能购买一件秒杀商品。如果用户成功秒杀到某商品,则显示订单支付页面,如图 9-7 所示。为突出重点,本章对于订单支付功能的实现不做介绍。

图 9-7 订单支付页面

9.2.2 实现过程

一个商品秒杀系统的基础功能是商品售卖。因此,本节讲解一个简单的商品售卖系统的开发过程。注:本节只讲解部分关键代码,完整代码见本书配套资源。

1. 创建数据库及数据表

在 MySQL 中创建名为 seckill 的数据库，并创建与秒杀业务相关的表：秒杀商品表 tb_sk_goods、秒杀订单表 tb_sk_order、用户表 tb_user、商品表 tb_goods 和订单表 tb_order，创建表单的语句如下（这里给出部分创建表单的语句，完整代码见本书配套资源）：

```
CREATE TABLE `tb_sk_goods`(
    `id` BIGINT(20) NOT NULL AUTO_INCREMENT COMMENT '秒杀商品 ID',
    `goods_id` BIGINT(20) NOT NULL COMMENT '商品 ID',
    `seckill_price` DECIMAL(10,2) NOT NULL COMMENT '秒杀价格',
    `stock_count` INT(10) NOT NULL COMMENT '库存数量',
    `start_date` datetime NOT NULL COMMENT '秒杀开始时间',
    `end_date` datetime NOT NULL COMMENT '秒杀结束时间',
    PRIMARY KEY(`id`) )ENGINE = INNODB AUTO_INCREMENT = 1 DEFAULT CHARSET = utf8mb4 COMMENT
'秒杀商品表';
```

2. 创建项目并完成 Spring Boot 的相关设置

创建一个名为 seckill 的 Spring Boot 项目，并配置相关依赖。相关 pom. xml 文件的内容见本书配套资源，此处略。

创建项目后，在 resources 目录下编辑 application. yml 文件，完成 Spring Boot 的相关设置，内容如文件 9-1 所示。

【文件 9-1】 application. yml

```
1   spring:
2     thymeleaf:
3       cache: false
4     datasource:
5       driver - class - name: com.mysql.jdbc.Driver
6       url: jdbc:mysql://localhost:3306/seckill?userUnicode = true
7               &characterEncoding = UTF - 8
8       username: root
9       password: 1234
10      hikari:
11        pool - name: secPool
12        minimum - idle: 5
13        idle - timeout: 180000
14        maximum - pool - size: 10
15        auto - commit: true
16        max - lifetime: 180000
17        connection - timeout: 30000
18        connection - test - query: SELECT 1
19    web:
20      resources:
21        add - mappings: true
22        cache:
23          cachecontrol:
24            max - age: 3600
25        chain:
26          cache: true
27          enabled: true
28          compressed: true
29          html - application - cache: true
```

```
30          static-locations: classpath:/static/
31    mybatis-plus:
32      mapper-locations: classpath*:/mapper/*Mapper.xml
33      type-aliases-package: com.example.seckill.pojo
34    logging:
35      com.example.seckill.mapper: WARN
```

如文件9-1所示，第3行关闭了thymeleaf的模板缓存；第4行开始配置数据源，项目中使用了hikari连接池，第11～18行配置了hikari连接池的相关参数；第19行开始设置Spring Boot的Web资源配置；第21行启动默认的静态资源处理机制，该项默认值true；第22行开启静态资源缓存，并设置响应内容在缓存中保存的最长时间，单位：秒；第25～27行配置资源链并开启资源链缓存，有关资源链的相关知识可参考Spring Boot的相关文档；第28、29行分别启用压缩资源(如gzip文件)解析器和HTML 5应用缓存；第30行配置静态资源所在路径。第31～33行设置了MyBatis-Plus的相关参数。

3. 创建持久化类

随后，创建与数据库表单对应的持久化类。下面给出与tb_sk_goods表单对应的持久化类，它代表参与秒杀售卖活动的商品，代码如文件9-2所示。

【文件9-2】 TSecKillGoods.java

```
1    package com.example.seckill.pojo;
2    //import 部分略
3    @Data
4    @TableName("tb_sk_goods")
5    public class TSecKillGoods implements Serializable {
6
7        @TableId(value = "id", type = IdType.AUTO)
8        private Long id;
9        /** 商品 ID **/
10       private Long goodsId;
11       /** 秒杀价格 **/
12       private BigDecimal seckillPrice;
13       /** 库存数量 **/
14       private Integer stockCount;
15       /** 秒杀开始时间 **/
16       private LocalDateTime startDate;
17       /** 秒杀结束时间 **/
18       private LocalDateTime endDate;
19   }
```

4. 编写Mapper接口及其实现类

以操作tb_sk_goods表为例，首先创建Mapper接口，代码如文件9-3所示。

【文件9-3】 TSecKillGoodsMapper.java

```
1    package com.example.seckill.mapper;
2    import com.baomidou.mybatisplus.core.mapper.BaseMapper;
3    import com.example.seckill.pojo.TSecKillGoods;
4    public interface TSecKillGoodsMapper extends BaseMapper<TSecKillGoods> {
5    }
```

其中，TSecKillGoodsMapper 接口继承了 MyBatis-Plus 提供的 BaseMapper < T >接口，后者提供了针对特定表单的插入、修改、删除和查询操作。这样 TSecKillGoodsMapper 接口就完成了对 tb_sk_goods 表单的相关操作方法的定义。

随后，创建 TSecKillGoodsMapper 对应的 Mapper。在 resources 目录下创建一个名为 mapper 的文件夹，并在该文件夹中创建一个名为 TSecKillGoodsMapper. xml 的文件，代码如文件 9-4 所示。

【文件 9-4】 TSecKillGoodsMapper. xml

```
1   <?xml version = "1.0" encoding = "UTF - 8"?>
2   <! DOCTYPE mapper PUBLIC " - //mybatis.org//DTD Mapper 3.0//EN"
3       "http://mybatis.org/dtd/mybatis - 3 - mapper.dtd">
4   < mapper namespace = "com.example.seckill.mapper.TSecKillGoodsMapper">
5       <!-- 通用查询映射结果 -->
6       < resultMap id = "BaseResultMap"
7           type = "com.example.seckill.pojo.TSecKillGoods">
8           < id column = "id" property = "id"/>
9           < result column = "goods_id" property = "goodsId"/>
10          < result column = "seckill_price" property = "seckillPrice"/>
11          < result column = "stock_count" property = "stockCount"/>
12          < result column = "start_date" property = "startDate"/>
13          < result column = "end_date" property = "endDate"/>
14      </resultMap >
15  </mapper >
```

5. 编写 Service 接口及其实现类

接下来，可以创建 tb_sk_goods 表的服务接口（Service 接口），代码如文件 9-5 所示。

【文件 9-5】 ITSecKillGoodsService. java

```
1   package com.example.seckill.service;
2   import com.baomidou.mybatisplus.extension.service.IService;
3   import com.example.seckill.pojo.TSecKillGoods;
4   public interface ITSecKillGoodsService extends IService < TSecKillGoods > {
5   }
```

其中，ITSecKillGoodsService 接口继承了 MyBatis-Plus 提供的通用接口 IService。IService 接口是应用于 Service 组件的接口，其作用与 BaseMapper 接口类似，在 Service 层对数据库操作进行封装，并提供了批处理操作功能。

对应地，创建 ITSecKillGoodsService 接口的实现类，代码如文件 9-6 所示。

【文件 9-6】 TSecKillGoodsServiceImpl. java

```
1   package com.example.seckill.service.impl;
2   //此处省略了 import 部分
3   @Service
4   public class TSecKillGoodsServiceImpl extends
5       ServiceImpl < TSeckillGoodsMapper, TSecKillGoods >     implements
6       ITSecKillGoodsService {
7   }
```

6. 编写控制器

本步骤介绍完成秒杀功能的控制器,其余控制器代码见本书配套资源。秒杀控制器代码如文件 9-7 所示。

【文件 9-7】 SecKillController.java

```
1  package com.example.seckill.controller;
2  //此处省略 import 部分
3  @Controller
4  @RequestMapping("/secKill")
5  public class SecKillController {
6
7      @Autowired
8      private ITGoodsService goodsService;
9      @Autowired
10     private ITSecKillOrderService secKillOrderService;
11     @Autowired
12     private ITOrderService orderService;
13
14     @RequestMapping("/doSecKill")
15     public String doSecKill(Model model, TUser user, Long goodsId){
16         if(null == user)
17             return "login";
18         model.addAttribute("user",user);
19         GoodsVo goodsVo = goodsService.findGoodsVoByGoodsId(goodsId);
20         if(goods.getStockCount() < 1){
21             model.addAttribute("errmsg", "该商品已售完!");
22             return "secKillFail";
23         }
24
25         TSecKillOrder secKillOrder = secKillOrderService.getOne(
26                 new QueryWrapper<TSecKillOrder>()
27                   .eq("user_id",user.getId())
28                   .eq("goods_id",goodsId));
29         if(secKillOrder != null){
30             model.addAttribute("errmsg","每人限购一件,请勿重复购买");
31             return "secKillFail";
32         }
33         TOrder order = orderService.secKill(user,goods);
34         model.addAttribute("order",order);
35         model.addAttribute("goods",goods);
36         return "orderDetail";
37     }
38 }
```

如文件 9-7 所示,秒杀控制器在响应秒杀请求时,首先判断用户是否登录(第 16～18 行)。在确认用户已登录的情况下,根据商品标记(ID)查找秒杀商品库存量(第 19～23 行)。为了防止用户重复抢购,第 25～32 行进行了用户是否重复购买的判断。最后,执行秒杀操作,并返回操作结果(第 33～36 行)。其中,第 19 行的 findGoodsVoByGoodsId()方法在 TGoodsServiceImpl 类中定义如文件 9-8 所示。

【文件 9-8】 findGoodsVoByGoodsId()方法定义

```
1    public GoodsVo findGoodsVoByGoodsId(Long goodsId) {
2        return tGoodsMapper.findGoodsVoByGoodsId(goodsId);
3    }
```

第 33 行的 secKill()方法在 TOrderServiceImpl 类中的定义如文件 9-9 所示。

【文件 9-9】 secKill()方法定义

```
1    @Transactional
2    public TOrder secKill(TUser user, GoodsVo goodsVo) {
3        TSecKillGoods secKillGoods = secKillGoodsService.getOne(
4            new QueryWrapper<TSecKillGoods>()
5                .eq("goods_id", goodsVo.getId()));
6        boolean secKillGoodsResult = secKillGoodsService.update(
7            new UpdateWrapper<TSecKillGoods>()
8              .setSql("stock_count = " + "stock_count - 1")
9              .eq("goods_id", goodsVo.getId())
10             .gt("stock_count", 0)
11       );
12       if (!secKillGoodsResult) {
13           return null;
14       }
15
16       //生成订单
17       TOrder order = new TOrder();
18       order.setUserId(user.getId());
19       order.setGoodsId(goodsVo.getId());
20       order.setDeliveryAddrId(0L);
21       order.setGoodsName(goodsVo.getGoodsName());
22       order.setGoodsCount(1);
23       order.setGoodsPrice(secKillGoods.getSeckillPrice());
24       order.setOrderChannel(1);
25       order.setStatus(0);
26       order.setCreateDate(new Date());
27       orderMapper.insert(order);
28       //生成秒杀订单
29       TSecKillOrder tSeckillOrder = new TSecKillOrder();
30       tSeckillOrder.setUserId(user.getId());
31       tSeckillOrder.setOrderId(order.getId());
32       tSeckillOrder.setGoodsId(goodsVo.getId());
33       secKillOrderService.save(tSeckillOrder);
34       return order;
35   }
```

如文件 9-9 所示，在执行秒杀操作时，首先根据商品标记(ID)在秒杀商品表(tb_sk_goods)中查找对应的商品，并获取该商品的基本信息(第 3～5 行)。随后，在该商品库存大于零的情况下扣减库存(第 6～11 行)。最后，生成订单和秒杀订单(第 16～33 行)。

至此，一个简单的商品售卖系统已开发完毕。虽然这个简单的商品售卖系统可以完成商品售卖，但它与一个秒杀系统还是有很大的差距。具体体现在以下几点。

(1) 存在超卖问题。可以用性能测试工具 JMeter 对本系统进行性能测试，测试结果表明，用户的订单数会远远多于待售的商品数，并且秒杀商品的库存数为负即出现超卖问题。这种情况在秒杀系统中是不允许出现的，而且这样的结果也是不正确的。

(2) 性能不佳。本书写作时利用JMeter进行性能测试，测试机的软硬件环境为：CPU 2.3GHz，双核Intel Core i5，内存8GB，macOS 13.4，MySQL 5.7，JDK 11。JMeter线程组数为1000，Rump-Up时间为0，循环次数为10。性能测试结果表明，该系统吞吐量为785.9/s。

9.3 改进方案

针对9.2节开发的商品售卖系统存在的问题，分析其原因如下：

(1) 超卖问题。出现超卖的情况是因为文件9-7第33行调用secKill()方法时，没有进行线程同步。多个线程同时读取秒杀商品的库存数，并进行扣减。如果没有线程同步，必然造成库存量减为负数的情况。要进行线程同步，最容易想到的解决方案是将secKill()方法的调用改写为如下形式。

```
1    synchronized(this){
2        TOrder order = orderService.secKill(user,goods);
3    }
```

然而，这并不能从根本上解决问题，因为synchronized关键字所封装的代码块只能保证在单个Java虚拟机(Java Virtual Machine，JVM)上实现线程同步。工程上，秒杀系统不可能部署到单一JVM上，而是部署到如图9-3所示的网关集群和业务服务集群上。这样的话，秒杀服务的应用程序代码会在多个JVM上同时运行，synchronized关键字无法保证多个JVM上的线程同步，超卖问题还会出现。

在分布式环境下解决多线程同步问题，一般可采用分布式锁或队列。对于队列，一般采用阻塞队列(java.util.concurrent.BlockingQueue < E >)和Disruptor队列。对于分布式锁，一般采用MySQL、Redis、ZooKeeper和Chubby实现。

利用队列解决线程同步问题的核心思想就是把接收到的请求按顺序存放到队列中，消费者线程逐一从队列里取数据进行处理。在分布式高并发场景下，由于进队和出队存在时间开销，因此会导致商品少卖。关于使用队列解决线程同步问题，本书不再赘述，读者可以参阅相关资料。

利用MySQL来实现分布式锁的思路是利用主键或索引冲突实现互斥。向表中插入一行数据时，如果有相同的主键或索引，则引起主键或索引冲突，进而起到线程互斥的作用。由于操作数据库有一定的系统开销，并且关系数据库的读写性能普遍不高，因此这种做法仅适用于中小型项目，对于秒杀这种分布式高并发的应用场景并不适合。

利用Redis来实现分布式锁，主要是借助Redis提供的SETNX命令和EXPIRE命令(用法见1.4.1节)。SETNX命令的含义可解释为SET if Not Exists(如果不存在某个键，则将键值存入Redis)，该命令是原子性的，如果参数key不存在，则当前key设置成功，返回1；如果当前key已经存在，则设置当前key失败，返回0。EXPIRE命令则可以用来设置key的生存时间。这两个命令配合使用，相当于某线程用SETNX命令对key进行加锁。同时利用EXPIRE命令设定该key的生存时间，起到完成时解锁的作用。加锁成功后，其他线程不能访问该key。待线程处理完毕，等待该key生存时间归0，完成解锁。Redis分布式锁的工作过程如图9-8所示。

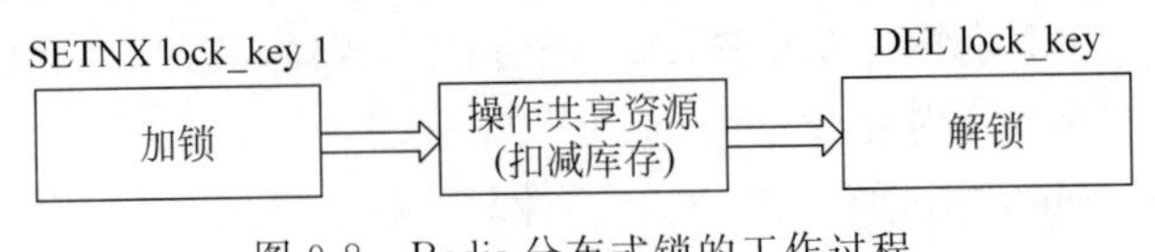

图9-8 Redis分布式锁的工作过程

借助Redis的命令虽然可以实现分布式锁，但在秒杀这类分布式高并发的应用场景下还是有很多无法解决的问题。而这些问题会造成分布式锁的失效。因此，在秒杀项目中要考虑使用更成熟的分布式锁机制。

（2）性能问题。从性能测试的结果来看，9.2节介绍的简单商品售卖系统的性能远远不能满足秒杀场景的要求。原因在于从读取商品库存、扣减库存到生成订单的过程，存在大量的数据库读写操作，而数据库读写操作会带来较大的系统开销，而且关系数据库的读写性能普遍不高。作为内存型数据库（或称作缓存），Redis具有良好的读写性能，并且能够应对大量的并发请求，这样的特点尤其适合用在秒杀场景中。为提升性能，应采用Redis作为秒杀系统数据的源。即初始库存数、扣减后的库存数全部存放在Redis中。

对于秒杀成功后生成的秒杀订单，可采用异步方式将订单保存到数据库。其基本思路为：秒杀成功后，在Redis中记录用户标识（用户ID）和商品标识（商品ID）。这样做的原因有二：一方面记录秒杀订单的关键信息；另一方面防止用户重复抢购。随后将上述两项内容通过消息队列（Message Queue，MQ）传递给消息接收器，由消息接收器生成订单并保存到MySQL数据库中，进而实现异步生成订单。本案例采用RabbitMQ作为消息队列，异步生成订单。

此外，对于秒杀活动来讲，参与秒杀的请求的数量远远高于售卖的商品的数量，只有少数请求可以成功购买到商品，而大多数请求都将无功而返。鉴于此情况，可以设置一个内存级别的变量，用来标记Redis中的库存数是否减为0。这样，参与秒杀的请求首先获取内存中的这个变量的值，如果该变量表明库存数不为0，则在Redis中扣减库存；否则通知客户端商品已售完。有了这个标记变量，可以减少大量的Redis访问，进一步提升性能。

综合上述分析，下面给出针对9.2节的简单商品售卖系统的改进方案，步骤如下。

1. 修改Spring Boot的配置文件

由于要利用Redis作为数据缓存，首先修改Spring Boot的配置文件application.yml，增加Redis及RabbitMQ的相关配置。需要增加的代码如文件9-10所示。同时，要在pom.xml文件中加入Redis及RabbitMQ相关依赖，具体内容见本书配套资源，此处略。

【文件9-10】 application.yml中增加的相关配置

```
1    redis:
2      host: 127.0.0.1
3      port: 6379
4      database: 0
5      timeout: 10000ms
6      lettuce:
7        pool:
8          max-active: 8
9          max-wait: 10000ms
```

```
10          max-idle: 200
11          min-idle: 5
12    rabbitmq:
13      host: 127.0.0.1
14      username: guest
15      password: guest
16      virtual-host: /
17      port: 5672
18      listener:
19        simple:
20          concurrency: 10
21          max-concurrency: 10
22          prefetch: 1
23          auto-startup: true
24          default-requeue-rejected: true
25      template:
26        retry:
27          enabled: true
28          initial-interval: 1000ms
29          max-attempts: 3
30          max-interval: 10000ms
31          multiplier: 1
```

2. 配置 RedisTemplate

随后，创建一个名为 RedisConfig.java 的文件，用于配置 RedisTemplate。代码如文件 9-11 所示。

【文件 9-11】 RedisConfig.java

```
1   package com.example.seckill.config;
2
3   @Configuration
4   public class RedisConfig {
5       @Bean
6       public RedisTemplate<String,Object> redisTemplate(
7               RedisConnectionFactory redisConnectionFactory) {
8         RedisTemplate<String, Object> redisTemplate = new RedisTemplate<>();
9         redisTemplate.setKeySerializer(RedisSerializer.string());
10        redisTemplate.setValueSerializer(RedisSerializer.json());
11        redisTemplate.setHashKeySerializer(new StringRedisSerializer());
12        redisTemplate.setHashValueSerializer(new
13          GenericJackson2JsonRedisSerializer());
14        redisTemplate.setConnectionFactory(redisConnectionFactory);
15        return redisTemplate;
16      }
17  }
```

3. 配置 RabbitMQ

本案例使用 RabbitMQ 完成秒杀订单的异步生成。因此，要在 Spring Boot 中配置 RabbitMQ 的消息队列和交换机，并将消息队列和交换机进行绑定。代码如文件 9-12 所示。

【文件 9-12】 RabbitMQConfig.java

```
1   package com.example.seckill.config;
2
3   //import 部分略
4   @Configuration
5   public class RabbitMQConfig {
6       private static final String SEC_KILL_QUEUE = "secKillQueue";
7       private static final String EXCHANGE = "secKillExchange";
8
9       @Bean
10      public Queue secKillQueue(){
11          return new Queue(SEC_KILL_QUEUE);
12      }
13
14      @Bean
15      public TopicExchange secKillExchange(){
16          return new TopicExchange(EXCHANGE);
17      }
18
19      @Bean
20      public Binding binding(){
21          return BindingBuilder.bind(secKillQueue())
22                  .to(secKillExchange()).with("secKill.#");
23      }
24  }
```

4. 创建异步消息发送端和接收端

在配置了RabbitMQ的消息队列和交换机后，还要创建消息的发送端和接收端。文件9-13和文件9-14分别给出了发送端和接收端的代码。

【文件 9-13】 MQSender.java

```
1   package com.example.seckill.rabbitmq;
2   //import 部分略
3   @Service
4   public class MQSender {
5       @Autowired
6       private RabbitTemplate rabbitTemplate;
7
8       public void sendSecKillMessage(String message){
9           rabbitTemplate.convertAndSend(
10                  "secKillExchange","secKill.message",message);
11      }
12  }
```

如文件9-13所示，第5、6行注入RabbitTemplate。RabbitTemplate是Spring-AMQP依赖提供的RabbitMQ消息模板，开发人员只需要使用Java代码来对RabbitTemplate进行配置，就可以将应用程序中的数据发送到RabbitMQ服务器中。第8行提供了sendSecKillMessage()方法用于发送消息。发送消息时，首先指定交换机，此处（第10行）指定的交换机secKillExchange就是文件9-12中第7行及第14～17行定义过的交换机；随后指定消息路由键secKill.message，这个路由键与文件9-12第22行定义的路由模式secKill.#匹配。

【文件 9-14】 MQReceiver.java

```
1    package com.example.seckill.rabbitmq;
2    //import 部分略
3    @Service
4    public class MQReceiver {
5        @Autowired
6        private ITGoodsService goodsService;
7
8        @Autowired
9        private RedisTemplate<String,Object> redisTemplate;
10
11       @Autowired
12       private ITOrderService orderService;
13
14       @RabbitListener(queues = "secKillQueue")
15       public void receiveSecKillMessage(String message) {
16           SecKillMessage secKillMessage = JsonUtil.jsonStr2Object(message,
17       SecKillMessage.class);
18           Long goodsId = secKillMessage.getGoodsId();
19           TUser user = secKillMessage.getUser();
20           GoodsVo goodsVo = goodsService.findGoodsVoByGoodsId(goodsId);
21           if(goodsVo.getStockCount()<1){
22               return;
23           }
24           TSecKillOrder secKillOrder = cKillOrder)redisTemplate.opsForValue()
25                   .get("order:" + user.getId() + ":" + goodsId);
26           if(secKillOrder != null)
27               return;
28           orderService.secKill(user,goodsVo);
29       }
30   }
```

发送端和接收端处理的都是 String 类型的消息,即发送端将传输的对象转换为 JSON 格式的字符串,接收端收到字符串后再将消息还原为对象(文件 9-14 的第 16、17 行)。消息中含有的两项关键内容为用户标识(用户 ID)和商品标识(商品 ID),可根据这两项内容创建订单。在创建订单前,先检查数据库中的实际库存,如果有足够库存,则可以继续创建订单(第 20～23 行)。第 24～27 行则检查用户是否已创建订单。如果没有创建订单,则在 MySQL 的订单表中创建用户订单(第 28 行)。其中,secKill()方法的定义参见文件 9-9。秒杀成功后,由于利用 Redis 保存秒杀订单,因此在文件 9-9 定义的 secKill()方法的第 33、34 行之间加入下述代码。

```
redisTemplate.opsForValue().set("order:" + user.getId() + ":" +
goodsVo.getId(),tSeckillOrder);
```

5. 缓存预热

如前面的分析所述,为提升性能,秒杀活动进行期间相关数据全部存放在 Redis 中。因此,在应用程序启动并完成配置后,要将相关数据存放到 Redis 中,这一过程称为缓存预热。同时,为了提升 Redis 的可用性,防止因为 Redis 服务器宕机造成秒杀活动失败,应采用 Redis 集群代替单机 Redis。本案例采用 7.3.3 节提出的 Redis 分片集群方案,将秒杀商品的初始库存存入Redis分片集群。此时，要修改文件9-7所示的SecKillController控制器代码。SecKillController类要实现InitializingBean接口（全限定名为org.springframework.beans.factory.InitializingBean），随后在该控制器内增加下面的方法，如文件9-15所示。

【文件9-15】 文件9-7增加的方法

```
1     public void afterPropertiesSet() {
2         List<GoodsVo> list = goodsService.findGoodsVo();
3         if(CollectionUtils.isEmpty(list))
4            return;
5         list.forEach(goodsVo -> {
6            emptyStockMap.put(goodsVo.getId(),false);
7            redisTemplate.opsForValue().set("secKillGoods:" + goodsVo.getId(),
8            goodsVo.getStockCount());
9         });
10    }
```

如文件9-15所示，第2行通过SQL语句从MySQL中获取秒杀商品的基础信息（包括秒杀价格、秒杀商品数量、秒杀活动开始和结束时间）。在获取到相关的商品信息后，第5～8行将获取到的秒杀商品库存信息以<商品ID:库存数量>的形式存入Redis，完成缓存预热。其中，第6行将内存标记emptyStockMap设置为false（表示商品尚未售完）。

6. 扣减库存，生成订单

完成缓存预热后，参与秒杀活动的商品库存已存入Redis。修改文件9-7的控制器代码（SecKillController），完成扣减库存和生成订单的任务，代码如文件9-16所示。

【文件9-16】 SecKillController.java

```
1     package com.example.seckill.controller;
2     //import 部分略
3     @Controller
4     @RequestMapping("/secKill")
5     public class SecKillController implements InitializingBean {
6         @Autowired
7         private RedisTemplate<String,Object> redisTemplate;
8
9         @Autowired
10        private MQSender mqSender;
11
12        //库存的内存标记
13        private Map<Long,Boolean> emptyStockMap = new HashMap<>();
14
15        @RequestMapping(value = "/doSecKill",method = RequestMethod.POST)
16        @ResponseBody
17        public String doSecKill(Model model, TUser user, Long goodsId){
18            if(null == user)
19                return "login";
20            TSecKillOrder secKillOrder = (TSecKillOrder)redisTemplate
21                 .opsForValue().get("order:" + user.getId() + ":" + goodsId);
22            if(secKillOrder != null)
23                return "请勿重复下单";
24
```

```
25          //通过内存标记减少 Redis 访问
26          if(emptyStockMap.get(goodsId))
27              return "商品已售完";
28          Long stock = redisTemplate.opsForValue()
29                  .decrement("secKillGoods:" + goodsId);
30          if(stock == 0L) {
31              emptyStockMap.put(goodsId,true);
32              return "商品已售完";
33          }
34          SecKillMessage secKillMessage = new SecKillMessage(user,goodsId);
35           mqSender.sendSecKillMessage(
36                  JsonUtil.object2JsonStr(secKillMessage));
37          return "success";
38      }
39
40      //缓存预热,商品数量加入 Redis
41      @Override
42      public void afterPropertiesSet() {
43          //见文件 9-13
44      }
45  }
```

如文件 9-16 所示，第 20～23 行检查用户是否重复下单，第 26、27 行检查商品库存的内存标记，如果该内存标记为 true，则说明对应的商品已售完，程序无须访问 Redis，这样可以进一步提升系统性能。第 28、29 行执行扣减库存操作。如果 Redis 中记录的某商品的库存减为 0，则设置相应的内存标记（第 30～33 行）。在用户成功抢购到商品后，将用户标识和商品标识封装为 SecKillMessage 对象，并将该对象转换为 JSON 字符串，利用 RabbitMQ 传递给消息接收端完成创建订单的任务（第 34～37 行）。其中 SecKillMessage 类的定义如文件 9-17 所示。

【文件 9-17】 SecKillMessage.java

```
1   package com.example.seckill.vo;
2   //import 部分略
3
4   @Data
5   @AllArgsConstructor
6   @NoArgsConstructor
7   public class SecKillMessage {
8       private TUser user;
9       private Long goodsId;
10  }
```

经过以上修改，引入了 Redis 进行库存初始化和库存扣减、引入 RabbitMQ 实现订单的异步创建，已基本解决了本节开头提出的性能问题。但多线程高并发场景下的超卖问题仍然没有得到解决。下面着重讨论如何解决超卖问题。

解决超卖问题还是要依靠分布式锁。本节给出三种分布式锁的解决方案：第一种方案，借助 Redisson 实现分布式锁；第二种方案，借助 Lua 脚本操作 Redis；第三种方案，利用 Spring 的 AOP 特性实现分布式锁。下面简要介绍这三种实现方案。

(1) Redisson 分布式锁。

Redisson 是一个分布式的 Redis 客户端,它提供了很多分布式服务。利用 Redisson 实现分布式锁很简单。

第一步,引入 Redisson 的依赖,代码如下。

```
1    <dependency>
2        <groupId>org.redisson</groupId>
3        <artifactId>redisson-spring-boot-starter</artifactId>
4        <version>选择合适版本</version>
5    </dependency>
```

第二步,创建 Redisson 的配置类,代码如文件 9-18 所示。

【文件 9-18】 RedissonConfig.java

```
1    package com.example.seckill.config;
2
3    import org.redisson.Redisson;
4    import org.redisson.config.Config;
5    import org.springframework.context.annotation.Configuration;
6    @Configuration
7    public class RedissonConfig {
8        public Redisson redisson(){
9            Config config = new Config();
10           config.useSingleServer().setAddress(
11               "redis://localhost:6379").setDatabase(0);
12           return (Redisson)Redisson.create(config);
13       }
14   }
```

文件 9-18 中,第 10 行调用 Config 类的 useSingleServer()方法,设置 Redisson 工作在 Redis 的单机模式下。事实上,Redisson 不仅支持 Redis 的单机模式,还支持 Redis 的各种集群模式(主从复制集群、哨兵集群、分片集群)。

第三步,利用 Redisson 实现分布式锁。需要使用分布式锁的是文件 9-16 所示的 SecKillController。首先在该类中加入对 Redisson 的引用。

```
1    @Autowired
2    private Redisson redisson
```

第四步,修改 SecKillController 的 doSecKill()方法,代码如下。

```
1    public RespBean doSecKill(Model model, TUser user, Long goodsId){
2        //见文件 9-16 的第 18~27 行
3        //获取 RLock 对象
4        RLock rLock = redisson.getLock("stock_lock");
5        try{
6            //加锁
7            rLock.lock();
8            Long stock = redisTemplate.opsForValue()
9                .decrement("secKillGoods:" + goodsId);
10           if(stock == 0L) {
```

```
11                  emptyStockMap.put(goodsId,true);
12                  return RespBean.error(RespBeanEnum.EMPTY_STOCK);
13              }
14              SecKillMessage secKillMessage = new SecKillMessage(user,goodsId);
15              mqSender.sendSecKillMessage(
16                  JsonUtil.object2JsonStr(secKillMessage));
17          } finally {
18              //解锁
19              rLock.unlock();
20          }
21          return "success";
22      }
```

Redisson 实现分布式锁的方式非常简单，其底层实现是依赖于 Lua 脚本。因此，可以自定义 Lua 脚本实现分布式锁。

（2）Lua 脚本操作 Redis。

当 Redis 执行 Lua 脚本时，Redis 会把 Lua 脚本作为一个整体并把它当作一个任务加入一个队列中，然后单线程按照队列的顺序依次执行这些任务。在执行过程中 Lua 脚本是不会被其他命令或请求打断的，因此可以保证每个任务的执行都是原子性的。使用 Lua 脚本，既可以将 Lua 代码保存在 Redis 服务器端，又可以把 Lua 脚本保存在客户端。简单起见，本案例将 Lua 脚本保存在客户端。

第一步，在项目的 resources 目录下创建一个名为 stock.lua 的 Lua 脚本文件，内容如文件 9-19 所示。

【文件 9-19】 stock.lua

```
1   if(redis.call('exists',KEYS[1]) == 1) then
2       local stock = tonumber(redis.call('get',KEYS[1]));
3       if(stock > 0) then
4           redis.call('incrby',KEYS[1], -1);
5           return stock;
6       end;
7           return -1;
8   end;
```

第二步，修改文件 9-16，加入 RedisScript 组件的引用。

```
1   @Autowired
2   private RedisScript<Long> script;
```

第三步，将扣减库存操作改由 Lua 脚本执行，即修改文件 9-16 的第 28、29 行。

```
1   Long stock = (Long)redisTemplate.execute(
2     script,Collections.singletonList("secKillGoods:" + goodsId),
3     Collections.EMPTY_LIST);
```

（3）Spring AOP。

前两种加锁方式都是在控制器层上实现的，可能显得不够优雅。还有一种方式就是在服务层（Service 层）实现加锁，步骤如下。

第一步，自定义锁注解，代码如文件 9-20 所示。

【文件 9-20】 ServiceLock. java

```
1   @Target({ElementType.PARAMETER, ElementType.METHOD})
2   @Retention(RetentionPolicy.RUNTIME)
3   @Documented
4   public  @interface ServiceLock {
5       String description()  default "";
6   }
```

第二步，定义切面类，代码如文件 9-21 所示。

【文件 9-21】 LockAspect. java

```
1    @Component
2    @Scope
3    @Aspect
4    @Order(1) //order 越小越先执行，但更重要的是最先执行的最后结束
5    public class LockAspect {
6        //互斥锁，参数默认为 false，不公平锁
7        private static  Lock lock = new ReentrantLock(true);
8
9        //Service 层切点
10       @Pointcut("@annotation(com.example.seckill.aop.ServiceLock)")
11       public void lockAspect() {
12       }
13
14       @Around("lockAspect()")
15       public  Object around(ProceedingJoinPoint joinPoint) {
16           lock.lock();
17           Object obj = null;
18           try {
19               obj = joinPoint.proceed();
20           } catch (Throwable e) {
21               e.printStackTrace();
22               throw new RuntimeException();
23           } finally{
24               lock.unlock();
25           }
26           return obj;
27       }
28   }
```

第三步，将文件 9-16 中定义的 doSecKill()方法封装到 Service 层，并添加相关注解。

```
1   @ServiceLock
2   @Transactional(rollbackFor = Exception.class)
3   public String doSecKill(Model model, TUser user, Long goodsId){
4   ...
5   }
```

第四步，创建一个新的控制器，调用 Service 层的 doSecKill()方法。

这种方法是利用 Spring 的 AOP 特性，在调用 Service 层的 doSecKill()方法时织入了

环绕通知。而环绕通知的作用就是在执行目标方法前加锁，在目标方法执行结束解锁，比较优雅地实现了分布式锁。

至此，商品秒杀系统开发完毕。

9.4 小结

本章利用 Spring Boot、Redis、RabbitMQ、MySQL 等开发了一个商品秒杀系统。秒杀系统是一个典型的分布式高并发应用场景。系统对性能、可用性的要求非常高。开发这一类应用系统，要时刻考虑高并发给开发工作带来的挑战。同时，要注重系统性能和可靠性的提升。而 Redis 作为内存型数据库，具有优异的并发读写性能，自然在秒杀系统的开发中得到广泛的应用。

开发秒杀系统最大的挑战就是发送给服务器的请求会在瞬间达到峰值。因此，秒杀系统面临的挑战有：

(1) 高并发。

秒杀的特点就是时间短，请求量极大。为了保证服务端正常运转，应该采用微服务的设计思想，将秒杀服务独立部署，为秒杀应用单独创建数据库及表单。并且，其服务网关应部署为集群架构，实现负载均衡。对于本章的案例，Redis 服务端采用的是 Redis 单机模式。在工程实践中，应将 Redis 服务端部署为 Redis 分片集群，并在缓存预热时，将秒杀商品的库存信息分别存放在 Redis 集群的不同节点上，从而实现负载均衡。使用 Redis 分片集群的意义在于提高 Redis 的可用性，避免因为单体 Redis 宕机而造成秒杀活动失败。

(2) 超卖。

造成超卖的原因在于秒杀场景下多线程同时扣减库存。因此，需要采用分布式锁进行线程同步。比较成熟的分布式锁的实现方式有 Redisson 分布式锁、Lua 脚本操作 Redis 及 Spring AOP 方式。

(3) 连接暴露。

本章围绕 Redis 应用介绍了一个简单的商品秒杀系统的开发。然而，对于真正的秒杀系统，本章介绍的内容还远远不够。通常的秒杀系统，在秒杀活动开始前，“秒杀”按钮都是置灰的，只有活动时间到了，才能点击。但是，如果提前获取了秒杀请求的接口，如通过浏览器的开发者模式查看网页源代码，那么某些恶意程序就可以在极短时间内将秒杀商品全部买下。这样的秒杀活动就失去了意义。为此，秒杀系统还应该隐藏秒杀接口地址，这部分内容与 Redis 应用无关，本章不再赘述。

(4) 其他问题。

从安全和限流角度考虑，秒杀系统还应配置用户登录验证码。利用验证码，一方面可以防止恶意程序频繁发送秒杀请求；另一方面可以起到限流的作用。

秒杀活动一般都是针对特定的商品。可以将商品列表页、商品详情页等以静态资源的形式存放在前端服务器上，这样可以减轻后端服务器的压力。

参 考 文 献

[1] 张云河,王硕. Redis 6 开发与实战[M]. 北京:人民邮电出版社,2021.
[2] 付磊,张益军. Redis 开发与运维[M]. 北京:机械工业出版社,2020.
[3] 金华,胡书敏. 基于 Docker 的 Redis 入门与实战[M]. 北京:机械工业出版社,2021.
[4] 谢乾坤. 左手 MongoDB,右手 Redis:从入门到商业实战[M]. 北京:电子工业出版社,2020.
[5] CARLSON J L. Redis 实战[M]. 黄健宏,译. 北京:人民邮电出版社,2015.

图书资源支持

感谢您一直以来对清华版图书的支持和爱护。为了配合本书的使用，本书提供配套的资源，有需求的读者请扫描下方的“书圈”微信公众号二维码，在图书专区下载，也可以拨打电话或发送电子邮件咨询。

如果您在使用本书的过程中遇到了什么问题，或者有相关图书出版计划，也请您发邮件告诉我们，以便我们更好地为您服务。

我们的联系方式：

清华大学出版社计算机与信息分社网站：https://www.shuimushuhui.com/

地　　址：北京市海淀区双清路学研大厦 A 座 714

邮　　编：100084

电　　话：010-83470236　010-83470237

客服邮箱：2301891038@qq.com

QQ：2301891038（请写明您的单位和姓名）

资源下载：关注公众号“书圈”下载配套资源。

资源下载、样书申请

书圈

图书案例

清华计算机学堂

观看课程直播